MANAGING COASTAL VULNERABILITY

MANAGING COASTAL VULNERABILITY

EDITED BY

LORAINE MCFADDEN
Flood Hazard Research Centre, Middlesex University, UK

ROBERT J. NICHOLLS
School of Civil Engineering and The Environment and the Tyndall Centre for Climate Change Research, University of Southampton, UK

and

EDMUND PENNING-ROWSELL
Flood Hazard Research Centre, Middlesex University, UK

ELSEVIER

Amsterdam • Boston • Heidelberg • London • New York • Oxford
Paris • San Diego • San Francisco • Singapore • Sydney • Tokyo

Elsevier
The Boulevard, Langford Lane, Kidlington, Oxford OX5 1GB, UK
Radarweg 29, PO Box 211, 1000 AE Amsterdam, The Netherlands

First edition 2007

Copyright © 2007 Elsevier Ltd. All rights reserved

No part of this publication may be reproduced, stored in a retrieval system
or transmitted in any form or by any means electronic, mechanical, photocopying,
recording or otherwise without the prior written permission of the publisher

Permissions may be sought directly from Elsevier's Science & Technology Rights
Department in Oxford, UK: phone (+44) (0) 1865 843830; fax (+44) (0) 1865 853333;
email: permissions@elsevier.com. Alternatively you can submit your request online by
visiting the Elsevier web site at http://elsevier.com/locate/permissions, and selecting
Obtaining permission to use Elsevier material

Notice
No responsibility is assumed by the publisher for any injury and/or damage to persons
or property as a matter of products liability, negligence or otherwise, or from any use
or operation of any methods, products, instructions or ideas contained in the material
herein. Because of rapid advances in the medical sciences, in particular, independent
verification of diagnoses and drug dosages should be made

British Library Cataloguing in Publication Data
A catalogue record for this book is available from the British Library

Library of Congress Cataloging-in-Publication Data
A catalog record for this book is available from the Library of Congress

ISBN-13: 978-0-08-044703-2
ISBN-10: 0-08-044703-1

For information on all Elsevier publications
visit our website at books.elsevier.com

Printed and bound in The Netherlands
07 08 09 10 11 10 9 8 7 6 5 4 3 2 1

Working together to grow
libraries in developing countries

www.elsevier.com | www.bookaid.org | www.sabre.org

ELSEVIER   BOOK AID International   Sabre Foundation

Contents

List of Figures	vii
List of Tables	xiii
Contributors	xvii
Acknowledgements	xix

1. Setting the Parameters: A Framework for Developing Cross-Cutting Perspectives of Vulnerability for Coastal Zone Management 1
   Loraine McFadden, Edmund Penning-Rowsell and Robert J. Nicholls

2. Vulnerability Analysis: A Useful Concept for Coastal Management? 15
   Loraine McFadden

3. More or Less than Words? Vulnerability as Discourse 29
   Colin Green and Edmund Penning-Rowsell

4. The Natural Resilience of Coastal Systems: Primary Concepts 45
   Colin D. Woodroffe

5. Integrating Knowledge for Assessing Coastal Vulnerability to Climate Change 61
   Jochen Hinkel and Richard J.T. Klein

6. The Vulnerability and Sustainability of Deltaic Coasts: The Case of the Ebro Delta, Spain 79
   Augustin Sánchez-Arcilla, Jose A. Jiménez and Herminia I. Valdemoro

7. Local Communities under Threat: Managed Realignment at Corton Village, Suffolk 97
   Sylvia Tunstall and Sue Tapsell

8. The Indian Ocean Tsunami: Local Resilience in Phuket   121
   John Handmer, Bronwyn Coate and Wei Choong

9. Vulnerability of the New York City Metropolitan Area to Coastal Hazards, Including Sea-Level Rise: Inferences for Urban Coastal Risk Management and Adaptation Policies   141
   Klaus Jacob, Vivien Gornitz and Cynthia Rosenzweig

10. Promoting Sustainable Resilience in Coastal Andhra Pradesh   159
    Peter Winchester, Marcel Marchand and Edmund Penning-Rowsell

11. Reducing the Vulnerability of Natural Coastal Systems: A UK Perspective   177
    Julian Orford, John Pethick and Loraine McFadden

12. Promoting Sustainability on Vulnerable Island Coasts: A Case Study of the Smaller Pacific Islands   195
    Patrick D. Nunn and Nobuo Mimura

13. Managing Coastal Vulnerability and Climate Change: A National to Global Perspective   223
    Robert J. Nicholls, Richard J.T. Klein and Richard S.J. Tol

14. Vulnerability and Beyond   243
    Loraine McFadden, Edmund Penning-Rowsell and Robert J. Nicholls

Subject Index   261

List of Figures

Figure 1.1:	Understanding vulnerability for coastal management: a simplified approach.	4
Figure 2.1:	Differences in recovery time from external forcing between physical and social systems.	19
Figure 2.2:	Developing regional- and hazard-specific methodologies for vulnerability assessment from generic assessment of coastal behaviour.	21
Figure 2.3:	Vulnerability analysis: restructuring the basic processes to promote new of 'better' functions.	25
Figure 3.1:	Conflict and choice.	31
Figure 3.2:	Signifier and signified.	35
Figure 3.3:	Words as metaphors.	36
Figure 3.4:	Words as seeds and elaborations.	37
Figure 3.5:	Towards a typology of definitions of vulnerability.	39
Figure 3.6:	A modified sustainable livelihood model of a household.	41
Figure 4.1:	Beach morphodynamic concepts illustrated for schematic beach between headlands.	47
Figure 4.2:	Three beaches adopting a different type of equilibrium (after Woodroffe, 2003).	48
Figure 4.3:	The concept of equilibrium.	49
Figure 4.4:	Temporal and spatial scales (based on Cowell & Thom, 1994, modified from Woodroffe, 2003), and significant processes on reefs.	50
Figure 4.5:	Coastal lagoon or barrier estuary, as found along coast of southeastern Australia.	53

viii List of Figures

Figure 4.6: Coastal lagoon showing coupling between beach states (as shown in Figure 4.1) and water level. 54

Figure 4.7: Schematic representation of different concepts of resilience. 55

Figure 4.8: Response of reef stratigraphy to sea-level change (after Woodroffe, 2003). 57

Figure 5.1: The DIVA development process. 68

Figure 5.2: Module linkages in the DIVA model. 69

Figure 6.1: The Ebro delta. 80

Figure 6.2: Subaerial surface changes in the Ebro delta from 1957 to 2000. 82

Figure 6.3: Area changes around the shoreline of the Ebro delta from 1957 to 2000. 83

Figure 6.4: The Ebro delta National Park area. 86

Figure 6.5: Overwash deposits in cultivated lands in the Marquesa beach after the impact of the November 2001 storm. *Source*: Spanish Ministry of Environment. 91

Figure 7.1: Map of the Corton seafront showing the location of the threatened holiday facilities and access to the seafront. 103

Figure 7.2: The small 'Wy Wurry' caravan park ironically the most immediately vulnerable of Corton's holiday facilities. 106

Figure 7.3: Caravans located at the cliff edge in another Corton caravan park threatened by coastal erosion. 106

Figure 7.4: Corton's coastal defences, sea wall, promenade and revetments collapsed due to undermining in the winter of 2000–2001. 107

Figure 7.5: Corton's coastal defences at high tide renewed in 2002–2003 with a narrow walkway on the sea wall and extensive rock cliff protection. 117

Figure 8.1: The location of the Southern Thailand island of Phuket. 123

Figure 8.2: The fate of dollars spent by overseas visitors to Phuket. 125

Figure 8.3: Number of domestic and international arrivals at Phuket Airport 2004–2005. 131

Figure 9.1: Map of Metropolitan East Coast Study Region with insert location. 142

Figure 9.2: Expected zones of storm surge flooding in lower Manhattan and parts of Brooklyn as a function of storm Level on the Saffir–Simpson Scale (SS 1–4). 143

Figure 9.3: Five models of projected sea level rise for the Battery at the southern tip of Manhattan, New York City. 145

List of Figures ix

Figure 9.4:   Map of the central portion of the MEC study area. Gray shading shows the areas at elevations below 3 m (10 ft to be exact) above the present mean sea level. 146

Figure 9.5:   Reduction in the 100-year recurrence period for three future decades and the five sea-level rise models shown in Figure 9.3. 147

Figure 9.6:   Current lowest critical elevations of facilities operated by the Port Authority of New York and New Jersey vs. changing storm elevations at these locations for surge recurrence periods of 10, 50 and 500 years between 2000 (baseline) and the 2090s. 148

Figure 10.1: Coastal Andhra Pradesh, India, showing the Divi Seema study area and the village survey transects. 160

Figure 10.2: Coastal vulnerability assessment model (Delft Hydraulics, 2003). 170

Figure 11.1: (a) The general trajectory of coastal forms as susceptibility increases and resilience decreases as a consequence. (b) An oblique view of the drift dominated spit developing westwards from the coastal exit of the river Sprey (Inverness, Scotland). (c) An oblique aerial view looking south over the swash-aligned and segmented beaches of Aberystwyth (Ceridigion, west Wales: photograph courtesy of Ceredigion County Council). 180

Figure 11.2: The tolerability of each coastal management policy option identified for the UK (DEFRA, 2001), as part of the preferred pathway over the next half-century. 182

Figure 11.3: A schematic view as to how an original wave-sediment cell of sediment source, transport corridor and sediment sink is disturbed and deflected by human intervention in pursuit of social and economic development centred upon the coastline. 183

Figure 11.4: (a) Artificial reprofiling of the gravel barrier at Cley (North Norfolk, England: photograph courtesy of John Pethick) to act as a flood defence scheme. (b) A comparison of unmanaged and managed barrier profiles at Cley (North Norfolk). 185

Figure 11.5: Structure (a) and location (b) of a Tan-y-bwlch gravel barrier (Aberystwyth, Ceredigion, west Wales). (c) Profile positions along the Tan-y-bwlch gravel barrier from which the annual rate of migration of index contours are shown in (d). (d) Annual rate of migration of specific index contours for profile positions along the Tan-y-bwlch gravel barrier. 188

Figure 11.6: The potential structure of UK future governance — consumerism domain into four scenarios related to scale of climate forcing and hence potential for coastal forcing. 191

x   List of Figures

Figure 12.1:   Map of the Pacific Islands regions showing the principal island groups mentioned in the text.   196

Figure 12.2:   (a) What remained of the K-Mart (supermarket) on Niue Island after Cyclone Heta in January 2004. (b) The place where Tuapa Church once stood on Niue Island, following the impact of Cyclone Heta in January 2004 (photo by Emani Lui, used with permission).   199

Figure 12.3:   (a) Navuti Village on Moturiki Island in central Fiji occupied a 40-m broad strip of low-lying coastal flat that lies 10–30 cm above mean high-tide level and is regularly flooded. (b) The buffer zone of vegetation between the ocean (on the right) and the village of Amuri (left) on Aitutaki Island in the Cook Islands (photo by Patrick Nunn).   200

Figure 12.4:   (a) Shoreline erosion manifested by fallen coconut palms along the back of the beach near Navitilevu Village on Naigani Island, central Fiji (photo by Patrick Nunn). (b) Coastal erosion at Nukui Village, Rewa Delta, Viti Levu Island, Fiji (photo by Nobuo Mimura). (c) Coastal erosion on Funafuti Atoll, Tuvalu (photo by Nobuo Mimura). (d) Beachrock on Ha'atafu Beach in Tongatapu Island, Tonga (photo by Nobuo Mimura).   202

Figure 12.5:   Map of Ovalau and Moturiki islands, central Fiji, showing the relative severity of shoreline erosion at every coastal settlement (after Nunn, 1999a, 2000a).   203

Figure 12.6:   Shoreline of a tropical Pacific Island resort in July 2003 has resulted in the removal of the sand covering the beachrock (darker area on right) and the limestone bedrock (centre) (photo by Patrick Nunn).   204

Figure 12.7:   Changes in settlement pattern on a typical smaller island in the tropical Pacific during the past 1200 years. (a) Settlement pattern at the start of the Little Climatic Optimum (Medieval Warm Period) about 1200 years BP (AD 750). (b) Settlement pattern towards the end of the Little Climatic Optimum about 700 years BP (AD 1250). (c) Settlement pattern during the early part of the Little Ice Age.   207

Figure 12.8:   (a) Piles (1 and 2) of *duva* (*Derris elliptica*) roots used for poisoning fish in a cleared area of mangrove swamps on the island Moturiki (central Fiji) (photo by Patrick Nunn). (b) Fragments of reef rock on sale on the side of the main highway, Lami Town, just outside Suva, the capital of Fiji (photo by Patrick Nunn).   210

Figure 12.9:   (a) Detail of a wall on Thulusdhoo Island, Maldives, to show the ways in which coral rock is utilized in such constructions when there are no other sources of hard rock available. (b) Remains of the seawall at Yadua Village, Viti Levu Island, Fiji (photo by Patrick Nunn).   212

Figure 12.10: Integrated system of natural coastal protection in the Pacific Islands. 216

Figure 13.1: Schematic view of the operation of the DIVA tool. 228

Figure 13.2: Summary of the sea-level rise impact assessment methodology within the FUND model 229

Figure 13.3: Sample results from DIVA: coastal flooding in 2000. (a) Flood plain population; (b) incidence of flooding at a regional scale; (c) incidence of flooding at a national scale and (d) incidence of flooding at a sub-national scale. 231

List of Tables

Table 5.1:	The vulnerability indicators of the IPCC methodology.	64
Table 5.2:	The modules of DIVA model.	73
Table 5.3:	Selected output of the DIVA model.	74
Table 6.1:	Estimation of the natural (function based) partial vulnerability index for the Ebro delta due to long-term coastline changes from 1957 to 2000.	87
Table 6.2:	Potential impact (ratio of affected surface to actual surface) of different units contributing to the deltaic natural function from 1957 to 2000 for different weighting scales.	88
Table 7.1:	Social and economic impacts of flooding and erosion on individuals and communities.	98
Table 7.2:	Intangible impacts of riverine and coastal flooding on individuals and communities.	99
Table 7.3:	Holiday facilities at Corton.	105
Table 7.4:	Descriptions of the options for adaptation.	109
Table 7.5:	Public response to adaptations to erosion through scheme options.	110
Table 7.6:	Priority for protecting different types of use of the coast.	114
Table 8.1:	Examples of informal sector occupations in Southern Thailand.	122
Table 8.2:	Size of the informal economy in Thailand in 2002.	125
Table 8.3:	Size of the informal sector across selected Nations.	128
Table 8.4:	The human cost of the tsunami by province in Southern Thailand.	130
Table 9.1:	Estimates of losses (in 2000-US$) in the MEC region for storms with shown surge heights.	149
Table 9.2:	Selected key institutions in the New York City Metropolitan area with a stake in coastal zone management.	153

xiv List of Tables

Table 9.3:	Stakeholder partners in the MEC project/coastal zone study.	154
Table 10.1:	The incidence and impact of severe and normal cyclones crossing the Andhra Pradesh coastline between 1949 and 1983 (Winchester, 1992, p. 7).	161
Table 10.2:	Density of population per square km in regions of Andhra Pradesh, 1971–2001.	162
Table 10.3:	Recovery between 1977 and 1997 in Krishna delta (Winchester, 2000).	164
Table 10.4:	The percentage of households in 2001 that perceived particular aspects of development as having the most negative impacts on their lives (Winchester & Penning-Rowsell, 2001; Table 3.1, Appendix IV).	166
Table 10.5:	The incidence of cyclones and extremes of climatic variation in coastal Andhra Pradesh (Winchester, 1992).	167
Table 10.6:	Development scenarios (Delft Hydraulics, 2003).	171
Table 10.7:	Coastal vulnerability model: results for Godavari Delta (Delft Hydraulics, 2003).	172
Table 11.1:	(Un)changing coastal vulnerability (V) through changing (Δ) susceptibility and resilience. Relative increases (+) or decreases (−) in susceptibility and/or resilience will cause a shift in vulnerability.	179
Table 11.2:	Varying manifest and latent coastal resilience outcomes based on engineered protection.	181
Table 11.3:	Estimated annual sediment availability between Flamborough Head and North Foreland (eastern England) and budget requirement for varying rates of sea-level rise.	186
Table 11.4:	Variable responses to possible future UK governance/values scenarios, affecting human approaches to coastal vulnerability through protection issues related to coastal barriers.	192
Table 12.1:	Characteristics of selected Pacific Island coasts.	197
Table 13.1:	Major impacts and potential adaptation responses to sea-level rise.	227
Table 13.2:	The four major physical impacts of sea-level rise, plus the adaptation approaches that are considered in the DIVA tool.	229
Table 13.3:	Global-mean sea-level rise scenarios for 1990–2100 for each SRES scenario.	232
Table 13.4:	Global population of the coastal flood plain (millions) in 2100. The results are independent of assumptions about adaptation.	233

Table 13.5: Global estimates of people flooded in 2100 (millions/year) assuming constant protection and economically optimum protection, respectively. 233

Table 13.6: Global estimates of people flooded in 2100 (millions/year) assuming constant protection and economically optimum protection, respectively. 234

Table 14.1: Five key areas where coastal vulnerability research in the social sciences would produce valuable results. 254

Table 14.2: Understanding coastal vulnerability: five key research topics in the environmental/engineering science areas. 256

Table 14.3: Understanding coastal vulnerability: five key research topics in integrated science. 257

Contributors

Wei Choong Centre for Risk and Community Safety, School of Mathematical and Geospatial Sciences, GPO Box 2476V, Melbourne, VIC 3001, Australia

Bronwyn Coate Centre for Risk and Community Safety, School of Mathematical and Geospatial Sciences, GPO Box 2476V, Melbourne, VIC 3001, Australia

Vivien Gornitz Center for Climate Systems Research, Columbia University, Mail Code 0205 and NASA Goddard Institute for Space Studies at Columbia University, Mail Code 0201, 2880 Broadway, New York, NY 10025, USA

Colin Green Flood Hazard Research Centre, Middlesex University, Queensway, Enfield EN3 4SF, UK

John Handmer Centre for Risk and Community Safety, School of Mathematical and Geospatial Sciences, GPO Box 2476V Melbourne, VIC 3001, Australia, and Flood Hazard Research Centre, Middlesex University, Queensway, Enfield EN3 4SF, UK

Jochen Hinkel Potsdam Institute for Climate Impact Research (PIK), P.O. Box 601203, D-14412 Potsdam, Germany

Klaus Jacob Lamont-Doherty Earth Observatory, Columbia University, P.O. Box 1000, Palisades, NY 10964, USA

José A. Jiménez Laboratori d'Enginyeria Marítima, ETSECCPB, Universitat Poltècnica de Catalunya, c/Jordi Girona 1-3, Campus Nord ed. D1, 08034 Barcelona, Spain

Richard J.T. Klein Potsdam Institute for Climate Impact Research (PIK), P.O. Box 601203, 14412 Potsdam, Germany

Marcel Marchand Marine and Coastal Management, WL | Delft Hydraulics, P.O. Box 177, 2600 MH Delft, The Netherlands

Nobuo Mimura Center for Water Environment Studies, Ibaraki University, Hitachi, Ibaraki 316-8511, Japan

Loraine McFadden Flood Hazard Research Centre, Middlesex University, Queensway, Enfield EN3 4SF, UK

Robert J. Nicholls School of Civil Engineering and the Environment and the Tyndall Centre for Climate Change Research, University of Southampton, Southampton SO17 1BJ, UK

Patrick D. Nunn Department of Geography, The University of the South Pacific, Suva, Fiji

Julian Orford School of Geography, Archaeology and Palaeoecology, Queen's University, Belfast BT7 1NN, UK

Edmund Penning-Rowsell Flood Hazard Research Centre, Middlesex University, Queensway, Enfield EN3 4SF, UK

John Pethick Independent Consultant, Beverley, East Yorks HU17 0DN, UK

Cynthia Rosenzweig Center for Climate Systems Research, Columbia University, Mail Code 0205 and NASA Goddard Institute for Space Studies at Columbia University, Mail Code 0201, 2880 Broadway, New York, NY 10025, USA

Agustin Sanchez-Arcilla Laboratori d'Enginyeria Marítima, ETSECCPB, Universitat Poltècnica de Catalunya, c/Jordi Girona 1-3, Campus Nord ed. D1, 08034 Barcelona, Spain

Sue Tapsell Flood Hazard Research Centre, Middlesex University, Queensway, Enfield EN3 4SF, UK

Richard S.J. Tol Centre for Marine and Climate Research, Hamburg University, Hamburg, Germany; Institute for Environmental Studies, Vrije Universiteit, Amsterdam, The Netherlands; and Centre for Integrated Study of the Human Dimensions of Global Change, Carnegie Mellon University, Pittsburgh, PA, USA

Sylvia Tunstall Flood Hazard Research Centre, Middlesex University, Queensway, Enfield, EN3 4SA, UK

Herminia I. Valdemoro Laboratori d'Enginyeria Marítima, ETSECCPB, Universitat Poltècnica de Catalunya, c/Jordi Girona 1-3, Campus Nord ed. D1, 08034 Barcelona, Spain

Colin D. Woodroffe School of Earth and Environmental Sciences, University of Wollongong, NSW 2522, Australia

Peter Winchester Flood Hazard Research Centre, Middlesex University, Queensway, Enfield EN3 4SF, UK

Acknowledgements

As editors we record our thanks to the authors, whose contributions comprise this volume. A number of individuals have also played an important role in the completion of this book. Susan Hanson has been pivotal to the submission of the volume with her diligence in handling the references lists and assistance in formatting author contributions. Ruth McFadden proofread a large number of the manuscripts and undertook a number of smaller jobs, which added up to a great measure of relief in the editing process. Yvette Brown, from the technical unit of Middlesex University, kindly undertook the role of formatting figures and tables throughout the volume. The Royal Society, London, provided the conference facilities for the initial exploratory workshop for this volume. Finally, our thanks go to the editorial team at Elsevier, especially to Joanna Scott who has been most supportive throughout the editing process.

LMF, RN, and EPR

Chapter 1

Setting the Parameters: A Framework for Developing Cross-Cutting Perspectives of Vulnerability for Coastal Zone Management

Loraine McFadden, Edmund Penning-Rowsell and Robert J. Nicholls

Introduction

The concept of vulnerability is often brought sharply into focus by disasters, whether induced from a natural forcing event or the direct result of human activity. The commencement of the 21st century has been marked by a series of catastrophic events and so issues surrounding 'vulnerability' are very much on the public, political and scientific agenda. The 2001 terrorist attack in New York City and Washington, DC embedded a deep realisation of stark vulnerabilities, which can define major urban areas. In the coastal zone, the devastating December 2004 Indian Ocean tsunami has raised important questions as to the vulnerability of coastal communities and the physical environment to such high-magnitude natural hazards. The impacts of Hurricane Katrina in Louisiana, Mississippi and Alabama, especially in New Orleans, in August 2005 have continued these concerns and raised a debate about coastal habitation in low-lying flood prone areas.

This book examines coastal vulnerability and so focuses on a particular range of pressures, responses and management approaches to vulnerable environments. At the boundary between the land and the sea, coastal systems occupy one of the most physically dynamic interfaces on Earth, encompassing a wide range of natural environments (McCarthy et al., 2001). Pressures on ecosystems within such environments are defined by an array of land-, river- and ocean-based drivers, and demands for goods and services from these systems are expected to increase for the foreseeable future (Millennium Ecosystem Assessment, 2005). Many economic sectors and major urban areas are located within the coastal zone. Indeed, the average population density within 100 km of the shoreline (112 people/km^2) is several times higher than the average global population density of 44 people/km^2 (Small & Nicholls, 2003). The coast also plays an important role in global transportation and the tourist industry. Coastal regions are

---

Managing Coastal Vulnerability
Copyright © 2007 by Elsevier Ltd.
All rights of reproduction in any form reserved.
ISBN: 0-08-044703-1

therefore complex, multi-functional systems with uniquely far-reaching and extensive conflicts of interests surrounding the use and management of coastal resources.

Alongside the important functional role of coastal systems are the significant hazards which impact these regions. The range of hazards faced by coastal communities can be extensive, and in addition to the hazards of more landward areas can include: surge and sea-water flooding, transmission of marine-related infectious diseases, extensive erosion and sedimentation hazards, hurricanes, tsunamis, oil spills and other technological-based hazards. Such threats underline the importance of understanding and managing the vulnerabilities of coastal environments and communities.

A brief introduction to the issues and questions that surround the concept of a vulnerable coastal zone, very quickly suggests that there are wide ranging spatial and temporal scales over which vulnerability can be considered. Some pressures on coastal systems are related to large-scale, high-magnitude/low-frequency events and have been a focus for disaster and global change research and management. However, the problems presented by vulnerable coasts and coastal communities can also be very much linked to the day-to-day management of local and regional-scale coastal behaviour. Coastal vulnerability research is therefore characterised by a wide range of challenging dimensions. Given the complex and debated nature of the concept, can vulnerability analysis bring useful insights to policies and strategies for managing the coastal system?

There seems to be a widespread consensus and concern about climate change and coasts and coastal nations have been urged to assess the vulnerability of their ecological and socio-economic systems to sea-level rise and other climate change impacts on the coastal environment (McCarthy et al., 2001). There have been considerable research interests surrounding the problems that lead to vulnerable coastlines and coastal communities and to methods of assessing vulnerability within the coastal zone (e.g. Capobianco et al., 1999; Thieler & Hammar-Klose, 1999, 2000; Pethick & Crooks, 2000; Nicholls, 2002, 2004; Adger et al., 2005). The wider disaster, climate change, human and food security literature that surrounds various forms of vulnerability to environment change is extensive. The context of sustainable development also provides a framework through which the type of real-world concerns (e.g., the WEHAB framework of the Johannesburg Summit, United Nations, 2002), which are often the impetuses for coastal vulnerability assessments, may be assessed. However, limited information exists as to how vulnerability can be actively reduced to promote the sustainable development and use of the coastal zone: that is, examining the potential impact and contribution of vulnerability analysis to coastal management.

This volume explicitly addresses, in this context, the question of the potential of the vulnerability concept to act as a basis for improving decision making in Coastal Zone Management. Its primary focus is therefore not on the problems that define vulnerable coastal systems *per se* or on methodologies or tools to quantify the vulnerabilities of coastal systems. Rather, the book sets out to explore the utility of vulnerability assessment as a tool for managing complex coastal systems.

Understanding specific coastal use issues and the processes that create vulnerable coastal systems remains essential to examining the implications of the concept for coastal zone management. Many chapters will discuss regional, scale and disciplinary-driven perspectives on the defining factors of vulnerable coastal systems. However, the context and challenge of this volume is to identify opportunities and barriers towards applying this knowledge in order to improve the basic status of coastal environments. In considering the effectiveness of the vulnerability concept as a tool for coastal zone management, it will seek to develop a series of cross-cutting (integrated) perspectives for developing sustainable coastal management strategies.

Themes within this Volume

Approaches to understanding the concept of vulnerability are contested. Recent advances in coupled social and ecological models of vulnerable systems (e.g. Walker et al., 2004; Adger et al., 2005) move towards resolving elements of conflict; while specific research communities (e.g., the climate change community) provide a broad-based platform from which commonalities of understanding and approaches can be nurtured. However differences remain in the semantics of vulnerability, and in the conceptual framework or world-views, which underpin the use of specific language in describing and modelling vulnerability. In turn, these differences in our understanding of vulnerability and vulnerability analysis result in a disparity in our responses to managing the coastal system. From a basic perspective, the coast is a contested environment, with many conflicts of interest surrounding the use of resources within coastal systems (Green & Penning-Rowsell, 1999). It follows that a value-loaded concept such as 'vulnerability' may be expected to reflect a bias towards a particular set(s) of ideals: it is not surprising that debate has surrounded the concept.

This volume does not attempt to resolve differences in approaches to the vulnerability term and each chapter is framed within the particular contributor's conceptualisation of vulnerability. This means that primary questions relating to the definition of vulnerability, and the metrics and approaches used to understand vulnerable environments and communities, have been left to the discretion of the contributors. Such a free-style approach to conceptualising vulnerability results in a high potential for complexity and diversity within the work, which raises a number of challenges for this volume. In describing a particular approach to 'vulnerability', the conceptual model could easily become the primary focus of the contributions. A cloud of concepts and conceptual frameworks can result in the real view of opportunities and barriers to the effective use of vulnerability being obscured. It is also more difficult to develop cross-cutting perspectives of opportunities and barriers for vulnerability reduction, on different and perhaps quite divergent models of understanding and analysing vulnerable environments.

With such challenges in view, this edited volume is loosely structured around a series of themes, based in the first instance on a simple relation:

Vulnerability = Impacts minus **effects of Adaptation** ($V = I - A$)

Unpicking this simple 'equation' leads to the following three themes:

Theme 1: Managing vulnerability through impact and adaptation responses.
Theme 2: Reducing impact on vulnerable coastal systems.
Theme 3: Enhancing adaptation in coastal environments and communities.

As already stressed, this framework is based within the context of decision making for coastal management, so that a *fourth theme*, that of the management of coastal environments, underpins and surrounds the approaches and discussions within the volume. 'Differences' in system behaviour and strategies for coastal management provides a *fifth and final theme*.

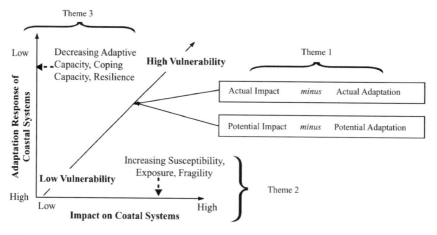

Figure 1.1: Understanding vulnerability for coastal management: a simplified approach.

In essence, the themes represent a series of pegs, on which each contributor can 'hang' their respective conceptualisations of vulnerability: thereby developing some consistency in approach without compromising important elements of diversity in the work. Figure 1.1 summarises the simple model of vulnerability on which this volume is structured. It highlights a series of concepts that are commonly associated with vulnerability assessments, and the basic relation between such concepts and the themed approached of this volume. Reducing the analysis to a series of building blocks (or themes) seeks to provide a common foundation through which diverse associations of concepts can be rooted. The model gives a consistent focus for developing ideas and methods for reducing the vulnerability of coastal systems to external forcing.

While this reductionist approach has obvious limitations, the common-denominating framework (i.e. $V = I - A$) enables basic similarities and differences across space and through time to be identified: both in the nature of vulnerable environments and, importantly, in the use of vulnerability analysis to improve the basic status of coastal systems. The aim of this volume is to move towards resolving some of the complexities in reducing the vulnerability of coastal systems, examining the contribution of the analysis to the integrated management of coastal systems. The themes therefore reflect a simplified, first-order view of a vulnerable coastal environment. However, they are grounded within the context of contemporary literature and within the perspective of developing guidelines and lessons to improve the sustainable management of complex coastal systems.

Theme 1: Managing Vulnerability through Impact and Adaptation Responses

There are two broad-based concepts that are most often used to describe the vulnerability of coastal systems. In the first instance is some idea of 'harm' (e.g. exposure, susceptibility, fragility) and on the other hand, recovery from the effects of external forcing (e.g. adaptive capacity, resilience and coping capacity). The framework of this volume ($V = I - A$) considers 'harm' within the context of 'impact' and 'recovery' in a broad-based theme of 'adaptation'. The basic premise on which the volume has been developed is that vulnerability can be reduced through: (1) decreasing the impact of external forcing on coastal systems and/or (2) enhancing the adaptive capacity of coastal environments and communities. The coastal zone management literature does not detail a conclusive relationship between vulnerability and these two basic responses of coastal systems. Rather, there is a measure of variability in approaches and relationships between the terms. Resilience, for example, has been considered a loose synonym for vulnerability, or vulnerability interchanged with the idea of 'harm' and so that adaptation is a direct response to increased vulnerability rather than increased impact on the system (Cutter, 2001; Vogel & O'Brien, 2004).

This volume considers the vulnerability of the coast as a general statement of the actual impact of a given external force on the coastal system, minus the effects of adaptation in response to that forcing within the total coastal environment. Based on this definition, vulnerability is an expression of the potential residual effects on a coastal society or environment given a particular hazard event, i.e. impacts over and above the effects of adaptation to the hazard. In turn, a vulnerable coastal system may be considered one in which the impacts of an external forcing event exceed mitigation or recovery that is reflected through the actual adaptation policies for the region. The emphasis of this perspective is the end-state of a system, after a (or a series of) impact and response cycle(s) to a hazardous event. The vulnerability of the coast is most often considered in such a context, with vulnerability assessments building an understanding of both present and future combinations of physical or socio-economic attributes, which define critical thresholds of impact and effective limits of adaptation responses.

However, in addition to the actual vulnerability of a coastal environment or community, it may be also possible to explore the potential vulnerability of the system. This perspective of vulnerability does not focus on the state of the physical or social environment i.e. those attributes of the physical and socio-economic systems that indicate critical limits of impact or adaptation. Rather, it explores processes which increase the likely impact of external forcing on the system and those which enhance the ability of the system to mitigate or absorb impacts on the coast. Such processes identify future changes to the actual thresholds of impact and response of a coastal system and therefore define potential vulnerability of the coast (McFadden, this volume).

Reducing the vulnerability of coastal environments and communities within this volume can relate to managing both the characteristics, and the underlying processes of coastal systems, to affect the impact of hazard forcing on coasts and the recovery of the coastal system.

Theme 2: Reducing Impact on Vulnerable Coastal Systems

The concept of vulnerability encapsulates the idea of a negative trend within the coastal zone, in the behaviour or the value of the coastal system. Therefore some notion of 'impact' is central to understanding the term. In managing the coast to reduce the vulnerability of coastal systems, a useful focus for 'impact' is the functionality of the coast: impact or harm may be summarised as some loss in the physical or socio-economic functional value of a coastal zone (McFadden, this volume). Given such an approach, attempts to decrease the vulnerability of coastal environments and communities would centre on diminishing the actual loss, or the potential towards a loss, of the functional value of the coast which may result from external forcing on the system. Actual impacts on the system reflect changes in the physical and socio-economic structure of the coast, e.g. the loss of coastal bluff by the order of x metres, or the inundation of x amount of homes. When exploring the potential towards impact, strategies for vulnerability reduction would focus on managing processes which increase the effects of hazard forcing on the system, e.g. physical processes which change (and de-stabilise) the composition of a soft coastline.

By developing a broad-based approach to this 'negative' component of the vulnerability equation, a range of perspectives on modelling the term can be accommodated (Figure 1.1). Some approaches to understanding impact, for example, particularly those focused on modelling the impacts of climate change, have centred on the physical susceptibility of coastal systems (Klein & Nicholls, 1999; Thieler & Hammar-Klose, 1999, 2000; McCarthy et al., 2001; Nicholls & Hoozemans, 2005). A high proportion of the vulnerability literature stresses the importance of modelling exposure to a specified hazard or range of hazard, where exposure examines the change of forcing and what is affected (e.g. Schiller et al., 2001; Smith et al., 2003; Turner et al., 2003; Schröter et al., 2004; Vogel & O'Brien, 2004; Adger et al., 2005). An exposure unit may reflect, for example, a region, population groups, community, ecosystem and country. However, it is a largely socially constructed phenomenon depending on where populations choose to live (or are forced to live), and how they construct their communities and livelihoods (Adger et al., 2004). The precise relationship between vulnerability and exposure is debated: is exposure a metric or a measure of vulnerability? Another dimension from the hazard-based literature is the relationship of fragility to vulnerability analysis. Fragility is a multi-dimensional function that reflects the fraction of the replacement value of an asset that is damaged when exposed to a specific hazard (Davidson et al., 2003; Chang, 2005). Through the use of 'impact' as a focal point for this discussion, similarities and differences in lessons for vulnerability reduction may be explored.

Characteristics of impacted coastal environments and communities, and processes that increase the effect of hazard forcing on coastal systems, will be explored in various levels of detail and complexity within this volume. An underlying theme is that some measure of the potential of the physical and socio-economic coastal sub-systems to be affected by the external forcing agent is central to understanding the vulnerability of the system. However, this impact, particularly within the physical environment, may be relatively fixed by broad-scale temporal and spatial processes. This means that applying such knowledge to the development of better management strategies for the coastal zone is challenging.

Theme 3: Enhancing Adaptation in Coastal Environments and Communities

Adaptation has become a strong element of vulnerability analysis. This reflects moves from general definitions and approaches to understanding vulnerability, towards operationalising the term for vulnerability assessments (e.g. Klein et al., 2001; Schiller et al., 2001; Yohe & Tol, 2002; Smith et al., 2003; Turner et al., 2003; Nicholls & Lowe, 2004; Vogel & O'Brien, 2004; Walker et al., 2004; Adger et al., 2005; Tol et al., in press). Actual adaptation measures are based on combinations of attributes that define the ranges across which a system can absorb external stress and perturbation. A potential adaptive response within a coastal system may be reflected through processes that enhance the resilience or adaptive capacity of a coastal system. The range of options for absorbing external forcing is often more extensive than those available to reduce the impacts of forcing on the coast. This means it is the adaptation response of coastal environments and communities that most often achieves the greatest returns in reducing the vulnerability of the system. As such, it is the 'positive face' of vulnerability analysis. However while adaptation is the general focus of a large number of vulnerability assessments, it is still relatively poorly understood from a strategic perspective, such as that relevant to coastal management.

While there is a strong expression in literature towards the role of adapting to external forcing, the relationship between the adaptive response of coastal systems and the vulnerability of the coast is relatively less distinct. As with the impact component of vulnerability, a large proportion of the discussion focuses on adaptation as a metric of vulnerability analysis. Some approaches consider adaptation obverse to vulnerability i.e. adaptation and vulnerability are two faces of the same coin. The adaptation-based literature is also divided in terms of the specific concepts used to describe the 'recovery' response (Figure 1.1). Many studies, for example, focus on the adaptive capacity of coastal systems. A range of definitions of adaptive capacity exist (McCarthy et al., 2001; Yohe & Tol, 2002; Smith et al., 2003; Adger et al., 2004), however it is generally described as the ability or capacity of a system to modify or change its characteristics or behaviour, so as to cope better with existing or anticipated external stress. Adaptive capacity is frequently cited in the context of human systems, as a societal-based concept describing active management of coastal systems. On the other hand, the concept of resilience has entered vulnerability analysis from ecology and largely maintains its association with the capacity of the ecological system to self-organise, although this is also often linked with societal response (Berkes, Colding, & Folke, 2003; Tompkins & Adger, 2004; Walker et al., 2004). The concept has evolved considerably since Hollings' (1973) seminal paper, however it retains much of its basic emphasis as the capacity of a system to absorb disturbance and reorganise while undergoing change so as essentially to retain the same function, structure and identity. A further dimension of adaptation is that of coping capacity: referring to a location-, group- and time-specific adaptive response of a system (Smith et al., 2003; Vogel & O'Brien, 2004).

This volume considers adaptation within the coastal zone in its widest sense and seeks to examine directions and challenges for enhancing the adaptation response across the coastal system. This means it incorporates perspectives on the capacity of both physical and socio-economic systems to absorb pressures towards adaptive change, reducing the vulnerability of the total coastal region.

Theme 4: Managing Coastal Systems

The aim of this volume is to explore vulnerability analysis as a tool for managing complex coastal systems, examining strategies that reduce the vulnerability of environments and communities. It is therefore important that the discussions are applied to the context of developing better strategies in decision making for coastal resource use and management.

The effectiveness of vulnerability analysis for coastal management relates to increasing scientific knowledge of total system behaviour of the coast (both physical and socio-economic systems) at both small–medium temporal (i.e. days — decades) and spatial scales (i.e. local — regional). However, the utility of vulnerability analysis as a tool for coastal management also depends on the effectiveness by which scientific understanding can be translated to management plans and strategies, improving the status of the coast (McFadden, this volume). This volume seeks to move the discussion on vulnerability analysis towards identifying a range of options and potential policy decisions for managing vulnerable and complex coastal systems to ensure a more sustainable coastal future.

Coastal Zone Management (CZM) is generally considered to reflect the definition, evolution, implementation and coordination of scientific procedures within the coastal zone to ensure its sustainable use and development (IPCC CZMS, 1992; Cicin-Sain, 1993; Kay & Alder, 1999). The complex, multi-dimensional behaviour of coastal systems to a wide range of stresses and perturbations requires that a strategic approach to managing the coastal environment is developed. The majority of the world's coasts, for example, have a legacy (and a future) of human occupation and are therefore the front line between their static socioeconomic constructs and dynamic, physical coastal systems (Carter, 1988; Hansom, 1988). Continual flooding, coastal erosion and loss of livelihood of coastal communities demonstrate the pressures faced by this unstable environment, and these problems appear to be increasing in intensity given accelerated global climate change. However, in addition to the challenges of understanding and modelling the dynamics of the physical environment, complex social processes underpin cultural perspectives of living at and managing the coast (de Groot & Orford, 2000). The discussion within this volume explores the challenges and opportunities in reducing the conflict of interests that define vulnerable coastal environments.

The specific focus of this volume is on integrated approaches to coastal management. As a result of a legacy of largely unsuccessful coastal management schemes, combined with increased pressures on the coastal zone, the 21st century has witnessed CZM becoming replaced by Integrated Coastal Zone Management (ICZM). Integrated management is characterised by a series of attributes, all of which are generally accepted within coastal management literature (e.g. Sorensen & McCreary, 1990; Vallega, 1993, 1999; Bower & Turner, 1997; Sorensen, 1997; European Commission, 1999; Kay & Alder, 1999; de Groot & Orford, 2000). These attributes can be considered as a function of one of two basic components of integration: (1) a perspective by which a coastal system is structured in an interdisciplinary way and (2) a commonality of purpose and approach between all stakeholders (e.g., scientists, policy makers, coastal managers and the public) within the coastal zone. Successful integration is based on the development of coastal management strategies from an agreement building process, which is defined by stakeholders and is underpinned by knowledge on the integrated behaviour of the coastal system (McFadden, in press).

The following chapters examine the degree to which an understanding of the vulnerabilities of coastal systems can move coastal management towards effective integrated management of coastal systems. The volume explores the range of dimensions characterising total system behaviour and seeks to identify cross-cutting perspectives, directions and policy decisions for managing vulnerable physical coastal environments and coastal communities.

Theme 5: 'Differences'

Cross-cutting themes and dimensions of applied coastal vulnerability analysis are important in developing strategic approaches to managing coastal zones. However, the complexity of processes, responses and drivers of change in coastal systems leads to a wide range of differential behaviour within and between particular coastal zones. This means that detailed recommendations for vulnerability reduction need to be context explicit. A high proportion of vulnerability literature stresses the necessity of developing specific context-based assessments of vulnerability (Kelly & Adger, 2000; Green, 2003; Turner et al., 2003; Adger et al., 2004; Vogel & O'Brien, 2004; Walker et al., 2004).

In the respect 'differences' are an important dimension of this volume. This theme refers to the range of variability within and between physical and social system behaviour, across spatial and temporal scales, and to the range of hazards to which coastal zones are subjected: ultimately reflecting a range of context-specific directions for reducing the vulnerability of particular coastal systems. Identifying key differences in system response allows specific components of a coastal zone to be targeted within a management context. A clearer understanding of the differences in the behaviour of vulnerable coastal environments may also improve the effectiveness of more strategic approaches in reducing coastal vulnerability — identifying the limitations and opportunities of broad-scale management strategies.

A Simple Route Map through this Volume

Exploring the Content of the Volume

In examining vulnerability as a tool for managing complex coastal systems, this volume draws from a wide range of perspectives and contexts of coastal environments. The book can essentially be divided into two constituent parts. The theoretical framing of the vulnerability term in the context of CZM is examined within (though not restricted to) the following four chapters, which together comprise a concepts-based section. McFadden (Chapter 2) argues that a system-based approach, focused on the dynamics of coastal behaviour, is of central importance in understanding vulnerability in complex coastal systems. The chapter suggests that the integrated functionality of the coast is a useful system-based framework for coastal management. Green and Penning-Rowsell (Chapter 3) and Woodroffe (Chapter 4) follow a dynamic approach to vulnerability with perspectives from social and physical sciences, respectively. Green and Penning-Rowsell suggest that the process of choice is central to defining vulnerability and that the usefulness of vulnerability analysis to coastal zone management is dependent on social constructions of the term. Woodroffe focuses on patterns, directions and rates of natural change that both coastal landforms and habitats undergo, stressing that the successful management of vulnerable coastlines depends on understanding natural processes of change. Hinkel and Klein (Chapter 5) use the example of a specific integrated-based project to consider the challenges of developing a domain-independent framework of vulnerability analysis. The chapter considers the process of communicating and integrating knowledge within vulnerability analysis.

The emphasis within the remainder of the book is the real-time application of ideas and approaches to managing vulnerable coastal environments. This 'applied' section begins with a consideration of the Ebro Delta as an example of a particularly vulnerable coastal system. Sanchez-Archilla et al. (Chapter 6) discuss a framework with which to assess deltaic vulnerability, examining the usefulness of vulnerability analysis as a tool for managing such environments. The focus then moves to a sociological perspective as Tunstall and Tapsell (Chapter 7) consider the challenge of strategic coastal management in the context of local community needs and perspectives of a functional coastal system. Using a coastal village in England as a case study, the chapter examines the complexities of managing the physical coastal environment against high levels of social vulnerability. Remaining within the theme of local communities Handmer et al. (Chapter 8) consider vulnerability in the context of local resilience in Phuket to the Indian Ocean tsunami. The chapter considers the gap between the rhetoric, and the reality, of securing local livelihoods as a critical component of management strategies towards reducing vulnerability.

In contrast to local sustainable livelihoods within developing regions, Klaus et al. (Chapter 9) discuss the capacities for adapting to coastal hazards within the New York City Metropolitan Area. The chapter gives some perspective of the challenges in translating vulnerability assessments of large vulnerable mega-cities into a coherent and sustainable approach to managing the population and resources of the region. Continuing at a regional scale, the focus of Winchester et al. (Chapter 10) is coastal Andhra Pradesh, India. Examining community resilience in a highly economically disadvantaged region, the discussion underlines the impact of relative affluence on adaptive capacity. Two chapters are based within a national context, giving a broad-scaled perspective on managing vulnerable coastal environments. Orford and his colleagues (Chapter 11) centre on approaches to enhancing sediment retention, and revitalising sediment pathways, as critical in reducing the physical vulnerability of the UK coastline. In a case study of the smaller Pacific Islands, Nunn and Mimura (Chapter 12) discuss coastal vulnerability within the wider integrated context of environmental and socio-economic change. In the final instance, Nicholls et al. (Chapter 13) reflect on international to global-scaled assessments of coastal vulnerabilities. Nicholls and his colleagues suggest that while coastal disasters are inevitable and adaptive responses complex, continuing progress on aggregated metrics of coastal vulnerability is making the policy choices on managing vulnerable environments somewhat clearer.

Examining the Dimensions of Coastal Behaviour

As the chapters explore opportunities and barriers to reducing the vulnerability of specific physical coastal environments and coastal communities, they address a range of dimensions in coastal behaviour.

One important dimension explored within this volume is thus that of spatial scale. A series of contributors examine local-scale characteristics, processes and approaches for reducing vulnerabilities (Sanchez-Archilla et al., Tunstall & Tapsell, and Handmer et al.) However, the volume extends this analysis through space to focus on broader-scale approaches to understanding the coast, from regional (Jacobs et al., Winchester et al.) through to national (Orford et al., Nunn & Mimura) and international perspectives on the vulnerability of coastal environments (Nicholls et al.). The variability of physical and socio-economic system response through time is also explored (McFadden and Woodroffe), with application for a low frequency, high magnitude event such as the Indian Ocean tsunami (Handmer et al.) and the long-term process of sea-level rise (Klaus et al., Nicholls et al.), through the day-to-day sustainable livelihoods of rural Indian farmers (Winchester et al.).

The contributions to this volume reflect specialist knowledge from a range of disciplines, including geomorphology (e.g., Woodroffe), coastal engineering (e.g., Sanchez-Arcilla et al.), sociology (e.g., Tunstall & Tapsell) and economics (e.g., Handmer et al.). Hence, a further dimension underpinning the volume is the multi-disciplinary approach to understanding and managing coastal environments. The specific coastal zones referenced throughout the volume represent a wide range of physical coastal types (e.g. delta, beach, cliffed coasts and coral reefs) and socio-economic settings (e.g. mega-city, island communities, small village), reflecting different conflicts of interest in coastal resource use and management.

Each key theme of vulnerability is explored in developed countries, i.e. the UK (Orford et al. and Tunstall & Tapsell), Spain (Sanchez-Arcilla et al.) and the US (Jacobs et al.) and in developing areas, i.e. India (Winchester et al.), Thailand (Handmer et al.) and the Pacific Islands (Nunn & Mimura). This means that broad-scale differences in issues, concerns and approaches to vulnerable coastal environments between these world-views can be examined.

The volume is structured to promote a broad view of vulnerable coastal environments. But specific perspectives on vulnerable environments and communities are also an important component of the book. However, the comprehensive framework of the volume gives important broad-scale perspectives on lessons and challenges for enhancing the utility of vulnerability analysis as a tool for managing coastal systems.

References

Adger, W. N., Brooks, N., Kelly, M., Bentham, G., Agnew, M., & Eriksen, S. (2004). *New indicators of vulnerability and adaptive capacity*. Final Project Report. Tyndall Project IT1.11. Norwich: Tyndall Centre for Climate Change Research, University of East Anglia.

Adger, W. N., Arnell, N. W., & Tompkins, E. L. (2005). Successful adaptation to climate change across the scales. *Global Environmental Change – Human and Policy Dimensions*, 15(2), 77–86.

Berkes, F., Colding, J., & Folke, C. (2003). *Navigating social-ecological systems. Building resilience for complexity and change*. Cambridge: Cambridge University Press.

Bower, B. T., & Turner, R. K. (1997). *Characterising and analysing benefits from integrated coastal zone management*. Working Paper, GEC97-12. Centre for Social and Economic Research on the Global Environment (CSERGE).

Capobianco, M., DeVriend, H. J., Nicholls, R. J., & Stive, M. J. F. (1999). Coastal area impact and vulnerability assessment: The point of view of a morphodynamic modeller. *Journal of Coastal Research*, 15(3), 701–716.

Carter, R. W. G. (1988). *Coastal environments: An introduction to the physical, ecological and cultural systems of coastlines*. London: Academic Press.

Chang, L. (2005). *Hurricane wind risk assessment for Miami-Dade County, Florida: A consequence-based engineering (CBE) methodology*. Hazard Reduction and Recovery Centre. Department of Landscape Architecture and Urban Planning, Texas A&M University.

Cicin-Sain, B. (1993). Sustainable development and integrated coastal zone management. *Ocean and Coastal Management*, 21, 11–44.

Cutter, S. L. (2001). A research agenda for vulnerability science and environmental hazards. *Newsletter of the International Human Dimensions Programme on Global Environmental Change*, 2(1), 8–9.

Davidson, R. A., Zhao, H., & Kumar, V. (2003). Quantitative model to forecast changes in hurricane vulnerability of regional building inventory. *Journal of Infrastructure Systems*, 9(2), 55–64.

de Groot, T. M., & Orford, J. D. (2000). Implications for coastal zone management. In: D. Smith, S. Raper, S. Zerbini, & A. Sanchez-Archilla (Eds), *Sea level change and coastal processes (DG12)*. Luxembourg: The European Union.

European Commission. (1999). *Towards a European integrated coastal zone management (ICZM) strategy. General principles and policy options*. EU Demonstration Programme on Integrated Management in Coastal Zones 1997–1999. Directorates-General Environment, Nuclear Safety and Civil Protection, Fisheries, Regional Policies and Cohesion.

Green, C. (2003). Change, risk and uncertainty: Managing vulnerability to flooding. *Third annual DPRI-IIASA meeting, integrated disaster risk management: Coping with regional vulnerability*, 3–5 July, Kyoto, Japan.

Green, C. H., & Penning-Rowsell, E. C. (1999). Inherent conflicts at the coast. *Journal of Coastal Conservation*, 5, 153–162.

Hansom, J. D. (1988). *Coasts*. Cambridge: Cambridge University Press.

Holling, C. S. (1973). Resilience and stability of ecological systems. *Annual Review of Ecological Systems*, 4, 1–23.

IPCC CZMS. (1992). *Global climate change and the rising challenge of the sea. Report of the coastal zone management subgroup intergovernmental panel on climate change response strategies working group*. Intergovernmental Panel on Climate Change, United States Natural, Oceanic and Atmospheric Administration, United States Environmental Protection Agency.

Kay, R., & Alder, J. (1999). *Coastal planning and management*. London: E and F N Spon.

Kelly, P. M., & Adger, W. N. (2000). Theory and practice in assessing vulnerability to climate change and facilitating adaptation. *Climatic Change*, 47(4), 325–352.

Klein, R. J. T., & Nicholls, R. J. (1999). Assessment of coastal vulnerability to climate change. *Ambio*, 28(2), 182–187.

Klein, R. J. T., Nicholls, R. J., Ragoonaden, S., Capobianco, M., Aston, J., & Buckley, E. N. (2001). Technological options for adaptation to climate change in coastal zones. *Journal of Coastal Research*, 17(3), 531–543.

McCarthy, J. J., Osvaldo, F., Canziana, N. A., Dokken, D. J., & White, K. S. (Eds). (2001). *Climate change 2001: Impacts, adaptation and vulnerability. Contribution of the Working Group 11 to the 3rd assessment report of the intergovernmental panel on climate change (IPCC)*. Cambridge: Cambridge University Press.

McFadden, L. (in press). Governing coastal spaces: The case of disappearing science in integrated coastal zone management. *Coastal Management*.

Millennium Ecosystem Assessment. (2005). *Ecosystems and human well-being: Synthesis*. Washington, DC: Island Press.

Nicholls, R. J. (2002). Analysis of global impacts of sea-level rise: A case study of flooding. *Physics and Chemistry of the Earth*, 27(32–42), 1455–1466.

Nicholls, R. J. (2004). Global flooding and wetland loss in the 21st century: Changes under the SRES climate and socio-economic scenarios. *Global Environmental Change-Human and Policy Dimensions*, 14(1), 69–86.

Nicholls, R. J., & Hoozemans, F. M. J. (2005). Global vulnerability analysis. In: M. Schwartz (Ed.), *Encyclopedia of coastal science*. Netherlands: Kluwer.

Nicholls, R. J., & Lowe, J. A. (2004). Benefits of mitigation of climate change for coastal areas. *Global Environmental Change-Human and Policy Dimensions*, 14(3), 229–244.

Pethick, J., & Crooks, S. (2000). Development of a coastal vulnerability index: A geomorphological perspective. *Environmental Conservation*, 27(4), 359–367.

Schiller, A., de Sherbinin, A., Hsieh, W., & Pulsipher, A. (2001). The vulnerability of global cities to climate hazards. *Paper presented at the open meeting of the Human Dimensions of Global Environmental Change Research Community*, 4–5 October 2001, Rio de Janeiro.

Schröter, D., Metzger, M. J., Cramer, W., & Leemans, R. (2004). Vulnerability assessment — analysing the human–environment system in the face of global change. *The ESS Bulletin*, 2, 11–17.

Small, C. & Nicholls, R. J. (2003). A global analysis of human settlement in coastal zones. *Journal of Coastal Research*, 19(3), 584–599.

Smith, J. B., Klein, R. J. T., & Huq, S. (Eds). (2003). *Climate change, adaptive capacity and development*. London: Imperial College Press.

Sorensen, J. (1997). National and international efforts at integrated coastal management: Definitions, achievements and lessons. *Coastal Management*, 25, 3–41.

Sorensen, J., & McCreary, S. T. (1990). *Institutional arrangements for managing coastal resources and environments*. Narragansett: University of Rhode Island.

Thieler, E. R., & Hammar-Klose, E. S. (1999). *National assessment of coastal vulnerability to future sea-level rise: Preliminary results for the U.S. Atlantic Coast*. Open-File Report 99-593. U.S. Geological Survey.

Thieler, E. R., & Hammar-Klose, E. S. (2000). *National assessment of coastal vulnerability to future sea-level rise: Preliminary results for the U.S. Pacific Coast*. Open-File Report 00-178. U.S. Geological Survey.

Tol, R. S. J., Klein, R. J. T., & Nicholls, R. J. (forthcoming). Adaptation to sea level rise along Europe's coasts. *Journal of Coastal Research*.

Tompkins, E. L., & Adger, W. N. (2004). Does adaptive management of natural resources enhance resilience to climate change? *Ecology and Society*, 9, 2.

Turner, B. L., Kasperson, R. E., Matson, P., McCarthy, J. J., Corell, R. W., Christensen, L., Eckley, N., Kasperson, J. X., Luers, A., Martello, M. L., Polsky, C., Pulsipher, A., & Schiller, A. (2003). A framework for vulnerability analysis in sustainability science. *Proceedings of the National Academy of Sciences, USA*, 100, 14, 8074–8079.

United Nations. (2002). *WEHAB framework documents*. World Summit on Sustainable Development, 24 August–4 September 1992. Johannesburg. Available online — www.johannesburgsummit.org/html/documents/wehab_papers.html (last accessed 2nd March 2006).

Vallega, A. (1993). The regional scale of Integrated Coastal Area Management: The state of conceptual frameworks. *Coastal Zone '93, Proceedings of the eighth symposium on Coastal and Ocean Management*, July 19–23, New Orleans, American Society of Civil Engineers.

Vallega, A. (1999). *Fundamentals of integrated coastal management*. The GeoJournal Library 49. New York: Kluwer.

Vogel, C., & O'Brien, K. (2004). Vulnerability and global environmental change: rhetoric and reality. *AVISO - Information Bulletin on Global Environmental Change and Human Security*, Issue No.13/2004.

Yohe, G. W., & Tol, R. S. J. (2002). Indicators for social and economic coping capacity — Moving towards a working definition of adaptive capacity. *Global Environmental Change-Human and Policy Dimensions*, 12(1), 25–40.

Walker, B., Holling, C. S., Carpenter, S. R., & Kinzig, A. (2004). Resilience, adaptability and transformability in social–ecological systems. *Ecology and Society*, 9(2), 5. Available online- http://www.ecologyandsociety.org/vol9/iss2/art5 (last accessed 2nd March 2006).

Chapter 2

Vulnerability Analysis: A Useful Concept for Coastal Management?

Loraine McFadden

Introduction

The concept of 'vulnerability' in coastal zone management (CZM) is far from new, and vulnerability assessments are frequently advocated in the development of risk-based coastal management programmes. International recognition of the concept is most clearly demonstrated in the Intergovernmental Panel on Climate Change (IPCC) Common Methodology for the Assessment of Vulnerability to sea-level rise (The Common Methodology) (IPCC CZMS, 1992). A plethora of sub-national and national vulnerability assessments have subsequently followed the IPCC approach to assessing vulnerability (Nicholls, 1995; McFadden, 2001).

As highlighted in Chapter 1, there have been high levels of concern about the problems that lead to vulnerable coastlines and about finding methods of assessing this vulnerability within the coastal zone. This concern is based on the fact that coastal systems still experience intensive and sustained pressures from a range of driving forces and that these 'drivers' are likely to be operative for many decades to come (e.g. Evans et al., 2006). However, more limited attempts have been made to examine the effectiveness of the vulnerability concept for CZM, and how the application of vulnerability analysis can contribute to management policies and strategies within the coastal zone. Indeed it is not clear that tools such as vulnerability analysis are conceived and defined in an effective manner for understanding and managing complex coastal systems. This chapter focuses on these issues surrounding the conceptualisation of vulnerability and the usefulness of the concept in coastal management.

Many studies that have cited the vulnerability concept have not defined either the notion of vulnerability or a vulnerable environment (e.g. Cooper & McLaughlin, 1998; Capobianco et al., 1999; Bryan et al., 2001; Hammar-Klose et al., 2003, 2004). There is also disparity when considering the components that comprise an analysis of vulnerability within the coastal zone. Some authors have focused on a combination of the susceptibility of a coastal system minus the resilience of the zone as a reflection of vulnerability within the region (e.g. Sánchez-Arcilla et al., 1998; de la Vega Leinert & Nicholls, 2001). Past studies have used the concepts of risk and vulnerability interchangeably (e.g. Gornitz, 1990; Alexander, 1992), while the terms sensitivity and vulnerability have also been used in an interchangeable manner (Pethick & Crooks, 2000). This illuminates the *ad hoc* legacy of vulnerability analysis that has been delivered via CZM. It also suggests a lack of scientific rigour in the way the vulnerability concept has been cited and applied in coastal management. Recent developments in vulnerability analysis, particularly within the global environmental change community, have focused on the importance of adaptive capacity. In this instance, vulnerability is considered some function of the exposure, sensitivity and the adaptive capacity (or coping dimension) of a system (Adger, 2000; Smith et al., 2003; Vogel & O'Brien, 2004; Adger and Vincent, 2005; Yohe et al., 2006).

It has also been argued that concepts, theories and philosophies do not often lend themselves to scientific definition and as well as proving difficult, it may actually be disadvantageous to seek a generic definition for the vulnerability term (Green & Penning-Rowsell, this volume). However, it is important that issues surrounding semantics do not overshadow the clear development of vulnerability analysis as an approach for managing the coastal zone. For example, without a general standard of good practice in the use of the vulnerability concept, integration across vulnerability analyses becomes difficult, if not impossible, to achieve. A poorly defined analysis leads to both inefficiency and redundancy of the approach as a coastal management tool (Hinkel & Klein, this volume).

This chapter will suggest a series of guiding principles for the definition and use of the vulnerability concept for CZM. As identified in chapter one, CZM is considered to reflect the definition, evolution, implementation and coordination of scientific procedures within the coastal zone to ensure its sustainable use and development (Kay and Alder, 1999). There are important differences across space and through time in the vulnerability of a coastal system: the guidelines seek to reflect basic integrating characteristics which can apply across a wide range of physical and social environments. Emerging from a review of conceptual models within vulnerability literature, the principles aim to increasing the usefulness of vulnerability analysis as a tool for understanding the problems associated with managing coastal change. They present the basic argument that vulnerability analysis should be considered as a comprehensive process-based assessment, embracing the entire coastal system. The chapter focuses on the potential value of a vulnerability approach to the long-term development and management of coastal systems and how, within a systems framework, vulnerability analysis could become a more useful tool for preserving and adapting key functionalities of the coastal system.

Principles for Conceptualising Vulnerability within CZM

Increasing the sustainability of development and management within the coastal zone often relates to: (1) enhancing our understanding and characterisation of physical and socio-economic processes within the coastal system and (2) providing effective means by which this understanding can be translated to policy-making within a region. The idea of 'usefulness' within this chapter is characterised on the basis of these two simple parameters. The principles provide a general framework in which complex scientific knowledge on coastal behaviour can be combined to give an integrated perspective on coastal change: ultimately explored within this chapter in the expression of the 'functionality' of the system. Understanding change, particularly change in the provision of goods and services from a coastal system, is a primary issue for coastal management. It is therefore an important vehicle for sustainable decision-making on resource use and coastal development.

Vulnerability as a Trans-Disciplinary Perspective on Coastal Change

A foremost principle in the use of 'vulnerability' for coastal management is that it should integrate physical and socio-economic ideals, to become a trans-disciplinary concept. Many vulnerability studies within the coastal zone have been based on a long heritage of traditional physical or social scientific viewpoints on the nature of change and the value of resources.

From the social science perspective, while considered a multi-dimensional concept, vulnerability is primarily conditioned by past, current and future populations and settlement patterns combined with the aggregated and per capita economic wealth of the region. In such approaches, the biophysical component, frequently considered as the exposure or measure of the hazard, is formally outside the definition of the term (Kelly & Adger, 2000). Governance by economic-based decision making has also been a specific feature of vulnerability analysis. The IPCC's common methodology vulnerability assessment, for example, is reduced in the final stage to a monetary value (McCarthy et al., 2001). A historical assessment shows that this economic approach to decision-making has dominated coastal management in general (Orford et al., this volume).

In engineering science, the concept of vulnerability is mostly linked to physical objects, e.g. houses, vehicles, so that in quantitative terms vulnerability is associated with the extent of structural harm or damage that results from an event (de Bruijn et al., in press). To the engineer, vulnerability analysis helps to promote structural integrity by addressing the way in which a structure is connected together. However, for the ecologist, vulnerability is related to biodiversity and functional redundancy. The concept of resilience in particular has entered vulnerability analysis from this subject area; introduced to emphasise the capacity of an ecosystem to bounce back to a reference state after disturbance, or maintain certain structures and functions despite increased forcing on the ecosystem (Holling, 1973; Turner et al., 2003).

Similarly, from a geomorphological perspective, vulnerability analysis is strongly related to relaxation periods; reflecting the time taken for a system to adjust morphologically to a change in energy input and regain a form of equilibrium. Examining the balance between the relaxation times and the return period of disturbing events determines the degree to which a system requires intervention before a loss of equilibrium occurs, and a new state is achieved (Pethick & Crooks, 2000). This ratio is considered to provide a critical measure of the manner in which coastal landforms respond to imposed changes, which determines the vulnerability of the coastal system. To use the vulnerability concept to its greatest potential in decision making, these different approaches need to be integrated into a common framework, achieving a more comprehensive assessment.

The need for inter- and trans-disciplinary research is becoming widely acknowledged and considerable discussions have surrounded multi-dimensional approaches to assessing vulnerability of coastal zones. This discussion has focussed on exploring the linkages between ecosystems and human societies: modelling vulnerability in the context of coupled socio-economic and ecological systems and the capacity of these systems to adapt to uncertainty and regenerate after disasters (Adger, 2000; Turner et al., 2003; Vogel & O'Brien, 2004; Walker et al., 2004; Adger et al., 2005).

Such discussions play an important role in defining the sustainability of coastal systems; however a central point must be raised: they reflect only two dimensions of the coastal landscape (i.e. societal and ecological systems). There has been limited debate as to how the vulnerability concept can be applied in the context of the total coastal system. Important questions such as: (1) the role of the physical state of the coastal zone as reflected in the morphological and sediment dynamics of the system i.e. the geomorphology of the coast (Woodroffe, this volume) and (2) the nature of interactions between these dynamics and socio-economic/ecological models have been inadequately (if at all) addressed. The fact remains that truly integrated approaches to modelling the problems and solutions to coastal change are still relatively few in number.

Understanding the vulnerability of the coast from such an integrated perspective is critical for CZM. The social construction of risk means that humankind will most frequently interpret a vulnerable environment when there is a threat to their socio-economic position through either direct or indirect loss. Thus, ultimately, it is 'memory' within the socio-economic system that drives the impact which humans have on the coastal zone. This memory can be defined as the time taken for the socio-economic system to adjust to an external forcing event. However, if memory within the physical system – the time taken for the ecological or geomorphological system to adapt to change and regain equilibrium – continually exceeds socio-economic memory, then society essentially has no gauge as to the forcing impact or the 'true' vulnerability of the system (Figure 2.1).

If coastal vulnerability is constructed in a trans-disciplinary manner, as a potential for change that evaluates the response of social systems conterminously with the range of physical responses of coastal processes (e.g. ecological and geomorphic) then short-term decisions can become more sustainable in the long-term development of the coast.

Beyond the Context of Climate Forcing: the 'Drivers' of Coastal Change

To be an effective basis for policy making in coastal management, vulnerability analysis must assess the impacts on coastal systems of a range of forcing agents. The relevant literature shows that a large majority of coastal vulnerability studies are characterised by their sole application to sea-level rise and the related effects of climate change upon the coastal zone (e.g. Nicholls & Nimura, 1998; Kelly & Adger, 2000; Bryan et al., 2001; Smith et al., 2003; Li et al., 2004; Pruszak & Zawadzka, 2005). This may be related to the centrality of the IPCC 'Common Methodology', where the overriding problem was viewed as sea-level rise and its impact on coastal resources. However, the value of the concept must be realised beyond the context of climate change. Although a major forcing agent within the coastal zone that must be accountable within models of vulnerability, climatic variation *per se* is not the sole driver of change experienced within the coastal zone. A range of drivers related to anthropogenic influences must also be considered if a

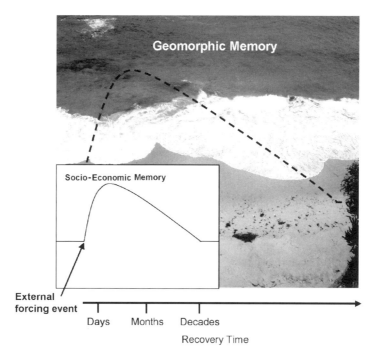

Figure 2.1: Differences in recovery time from external forcing between physical and social systems.

comprehensive tool is to be developed that is of significant value to CZM. Many of the world's open coasts and estuaries, for example, are extensively developed with high levels of population, property and infrastructure resulting in development and regeneration pressures being key drivers of change.

Recognising the importance of modelling coastal system response to a range of physical and socio-economic drivers is not a new phenomenon (de la Vega Leinert & Nicholls, 2001; McFadden, 2001). The Foresight Flooding and Coastal Defence Project run by the UK Office of Science and Technology is a key example of progress towards addressing this issue (Evans et al., 2004). Producing a long-term vision for the future of flood and coastal defence in the UK, the project focused on a wide range of drivers that may change the state of the flooding system; these included climate change, urbanisation, changing agricultural practices and rural land management.

The UK Foresight Project is an important example of a comprehensive assessment incorporating the context and impacts of change within fluvial and coastal systems. However, the bulk of vulnerability assessments for CZM do not facilitate such a broad-scale approach to understanding the drivers of coastal behaviour. Developing frameworks and approaches to modelling vulnerability which are embedded within a comprehensive analysis of the drivers of change within the system must become a goal for the CZM community.

Local Studies to Broad-Scale Assessments: Vulnerability across Spatial and Temporal Scales

The case studies which comprise the bulk of this book demonstrate a range of local and regional factors which condition the vulnerability of specific coastal systems to various forcing factors. Many vulnerability studies within the literature focus on fine-scale variability of coastal behaviour in response to environmental or socio-economic forcing. Given such variability, it is argued that the concept of vulnerability can only be meaningfully examined against a particular hazard or spatial scale (e.g. Cooper & McLaughlin, 1998; Kelly & Adger, 2000): indices being developed at a specified scale, for a specified risk against a range of management scenarios.

Multiple forcing stimuli and the complexities and dynamic nature of coastal behaviour do mean that scale, region and hazard-specific indicators of vulnerability are important in decision making: this is particularly true at the local management scale. However, restricting the concept in such a manner means that vulnerability must be redefined for every environmental perturbation at every scale of interest within the coastal zone. Given such demands, can the concept form the basis of a strategic approach to coastal management policy?

The value of broad-scale coastal analyses has been widely recognised (Nicholls et al., this volume; Nicholls et al., in press). Broad-scale vulnerability assessments can be useful as a tool for identifying sections of the coast that require further analysis or as a basis for regional planning and management guidance. Such analyses increase understanding of a wide range of potential coastal system behaviours given multiple sources of external forcing and can therefore play a particular role in policy making for CZM.

A broad-scale behavioural approach to vulnerability analysis identifies the different elements, which comprise the total coastal system, and develops an understanding of how these elements interact on a range of both spatial and temporal scales. Many uncertainties surround the broad-scale relationships between physical and human systems, feedback linkages between the two environments, and scaling issues (across space and in time), which need to be explored. Modelling this range of behaviour–response scenarios within a coastal system in a comprehensive vulnerability analysis illustrates the potential reaction of the total coastal environment given external forcing. Such a perspective can form a useful basis for a strategic framework in CZM. Regionally scaled methodologies based on particular hazards, or geographically defined behaviour of the system, links this large-scale understanding of the system to smaller-scaled sub-regional/local studies (Figure 2.2).

A comprehensive approach to assessing the response of the total coastal system (to the range of drivers of system change as highlighted in the previous section) could add significant value to vulnerability analysis for coastal management. Detailed, local response-driven models are critical; however the necessity of such analysis should not negate the value of broad-scale models. Vulnerability assessments which can examine the primary interactions, links and responses defining the behaviour of the total system when subjected to external forcing may play an important role in strategic, long-term and broad-scale management of the coast (McFadden, in press). Understanding the total behaviour of the system across spatial and temporal scales remains a fundamental problem for CZM. Given that vulnerability may be more highly linked to large-scale processes today than in the past

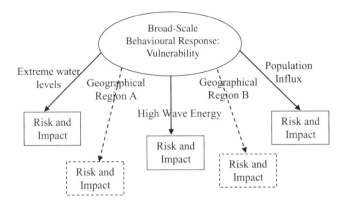

Figure 2.2: Developing regional- and hazard-specific methodologies for vulnerability assessment from a broad-scale assessment of coastal behaviour.

(Adger et al., 2005), such approaches may provide increasingly important insights for understanding and managing coastal change.

The Metrics of Vulnerability Analysis: Modelling the Primary Processes of Change

The first guiding principle of this chapter focussed, above, on vulnerability as a trans-disciplinary concept: the second and third principles stress the importance of a comprehensive approach to understanding the drivers and subsequent responses of the system to changes in external forcing. To be effective when understanding conflicts of interest within coastal environments, it is important that such perspectives are also framed within a strong process-based analysis of the coastal system. The metrics of vulnerability analysis should reflect the principal driving processes that define the behaviour of the coast (McFadden et al., 2006).

The exact label given to these metrics may actually be relatively insignificant, i.e. do we use the idea of resilience or adaptive capacity to describe negative feedback in coastal response? Physical systems have a capacity to absorb external forcing, producing a dynamic *status quo* within the environment. This means that physical systems are in principle self-organising, although their capacity can be and often is limited, either naturally or by human agency. In reality, many physical coastal systems are 'managed' to promote self-organisation. This concept of resilience is most frequently associated with physical coastal environments i.e. ecological or geomorphic-driven studies. The converse is true in the context of society, which largely must be managed to be resilient, a concept most frequently described as developing adaptive capacity. Both concepts (i.e. resilience and adaptive capacity) are used in different scientific communities with different conceptual and empirical backgrounds. However, despite these differences, the basic principle underpinning both concepts is essentially the same and reflects the dynamic (and managed?) response of the system when subjected to a disturbance (de Bruijn et al., in press).

It may be argued that rather than multiplying or re-defining the metrics and concepts of vulnerability analysis, more emphasis should be placed on understanding the interactions, differences and similarities between system processes which underpin existing approaches to conceptualising the term. It is important to achieve a greater understanding of the potential for, and the direction of change (both now and in the future), in the critical states or behaviour which defines vulnerable systems. Adding new dimensions (and metrics) of vulnerability *per se* may not necessarily improve the effectiveness of the analysis, if that analysis does not reflect the socio-economic and physical processes that drive change. Vulnerability analysis should therefore assimilate a more concrete understanding of the processes which define the behaviour of the system within the current threshold-driven approach to assessing vulnerability. Threshold analysis is of course implicitly rooted in system processes: any given system state represents a (dynamically) stable process environment. However, there is often little explicit link between the thresholds that define the vulnerability of a system and the processes (physical and social) which underpin this expression of coastal behaviour. It is really only when such a process-based framework is obtained that the metrics of vulnerability analysis become indicators of system change, rather than indicators of the static 'state of the coast'. Moving towards such a dynamically based vulnerability analysis is critical for the effective use of the concept in complex system management (Downing, 2003).

Terminology (i.e. the concepts in vulnerability analysis) is only important in that it gives insight into the decisions which must be made and carries shared meaning which enables stakeholders to communicate with each other (de Bruijn et al., in press). The key point is that in developing approaches to assessing vulnerability, the components of the analysis must give insight into the processes which define the behaviour of the total system.

Vulnerability as a Comprehensive, Systems-Based Analysis

The principles for conceptualising vulnerability within this chapter present a basic argument for the use of the term in CZM: scientists, policy makers and other stakeholders involved in managing the coast should consider vulnerability analysis as a comprehensive process-based assessment, embracing the entire coastal system. This section suggests an approach to adapting vulnerability as a comprehensive assessment of coastal behaviour: highlighting a functionality model as a useful vehicle for understanding the complex response of coastal systems. The section focuses on the potential value that such a systems-based approach to vulnerability analysis could add to the suite of CZM tools which currently exist for coastal managers.

Integrated System Functionality: Complimenting Existing Tools for CZM

A range of tools has been developed and accepted as instruments for decision making in CZM, for example, cost–benefit analysis and environmental impact assessment. The monetisation of coastal resources, through economics-based tools, provides a rational and important approach to managing the coastal zone: assessing economic costs and benefits of decisions and management options is a primary role within effective CZM. While tools such as cost-benefit analysis provide important information for coastal management, they do not attempt to assess non-use values and do not reflect the integrated behaviour of the coastal system. However such tools are also strongly biased towards system outputs or deliverables, with no real focus on the structures and processes necessary to deliver the relevant goods and services. A wetland may be valued, for example, on the basis of a series of preferred services to society e.g. bird habitat and storm protection. However, such valuations are not associated with the processes which create the wetland environment and subsequent functions that the system provides. An understanding of the relationship between socio-economic processes and highly valued cultural landscapes of the coast is another example of the importance of processed-based analysis to long-term management. By taking a basic black-box approach (i.e. focusing on the system outputs), society loses a powerful tool for preserving and adapting such key functionalities of the system.

Adapting 'vulnerability' on the basis of the functionality of the coastal system may be one vehicle whereby a systems-based understanding of the coastal zone could be translated to a tangible method for decision-making in CZM. The idea that the natural system provides functions for human existence is not new, nor is the link between system functions and processes central to under-pinning the value of environmental systems. However, the fact remains that many coastal management tools do not in practice realise the link between system deliverables and the sustainability of the structures and processes necessary to deliver goods and services. The network of beaches, wetlands and hard-rock coastlines at the coast provides a range of functions in both a socio-economic and natural capacity. A sandy beach, for example, is a buffer against wave attack to prevent flooding and land loss through erosion, as well as a recreational resource; coral reefs are effective coastal protection structures and sand dunes form natural bluffs and sand repositories from which sand may be extracted during storms. Wetlands are highly productive systems providing, for example, waste assimilation, flood protection, nursery areas for fisheries and habitats for wildfowl (McLean et al., 2001). If vulnerability assessments were tied to a view of the coast as an environment in which a range of functions are overlain, this may provide a platform from which the dynamics of system processes and thresholds of coastal behaviour could be explored and translated to policy-making for coastal management.

A 'functionality approach' would not focus primarily on the 'deliverables' but rather on the sustainability of ecosystem structures and processes necessary to deliver goods and services. It focuses attention on simple questions such as 'why does a sandy beach appear in this particular system?' or 'why do we associate a high cultural value with this specific landscape?' The buffering capacity of a sandy beach, for example, can be related to processes that control the morphological and sediment structure of the system. Cultural heritage and social sensitivity to particular landscapes are often driving forces in defining the aesthetic value of coastal environments. By examining the dynamics that create and maintain system behaviour, it assesses the ability of a system to maintain a range of functions through time. If 'vulnerability' was defined in two dimenions i.e. (1) the critical thresholds defining the limits of behaviour and (2) the capacity of the system to maintain that behaviour, it may afford the opportunity to explore and value processes in a manner that other tools such as Cost-Benefit Analysis and Multi-Criteria Analysis cannot allow.

Such an approach would require a greater understanding of the complexities between the adjustments of processes and form in coastal systems, as well as interactions between physical and socio-economic environments. This may be a challenging, but important, direction for coastal science within the 21st century.

Defined in the context of functionality, a coastal system can be considered vulnerable if its functions are easily threatened, and hence easily degraded such that its outputs are markedly lower. Such coastal functions are anthrogogenically referenced, e.g. decreasing vulnerability implies increasing stability of the costal environment in terms of its structure and functionality. Decreasing vulnerability is therefore associated with a positive outcome for humans whereby coastal stability is equated with support of human functionality, whereas increasing vulnerability identifies destabilised coastal environments and degraded human functionality (Orford et al., this volume). Such an approach is essentially anthropogenic in its basis: however CZM is itself anthropogenic driven — as a physical entity, there is no need for management within the coastal system.

A functionality perspective on vulnerability analysis is one approach which contributes to available knowledge of coastal behaviour for CZM. It provides some estimation of the integrated value of the coastal systems and links this assessment to the basic physical, social and economic processes which define the structure and functionality of the coast. Ultimately, functionality focuses on a dynamic systems approach to understanding the behaviour of the coast.

Challenges and Benefits From a Complex Functional-System Perspective on Vulnerability Analysis

A functionality approach to vulnerability analysis contributes to our understanding of processes which underpin values attributed to the coastal system. In this way, it provides a useful context for developing management policies and approaches: increasing our understanding of conflicts of interest that threaten the utility of the coastal zone. It is a potential tool for understanding the coast as a complex system defined by dynamic interactions of multiple processes (geomorphic, ecological, economic and societal) to a wide range of drivers of coastal change. By identifying areas along the coast where there is a high potential for sudden loss in the integrated functionality of a coastal system given external forcing, the particularly sensitive points in the coastal zone may be targeted for management to minimise the loss of the resource base.

Building vulnerability analysis on the integrated nature of processes that underpin the behaviour (and the functional value) of the total coastal system would allow the changing use of the coastal zone to be managed as part of a strategic policy: ensuring key values of the coast are not reduced. In this manner, vulnerability analysis can contribute to the potential long-term development of the coast, providing a framework within which the concept can be rooted as a basis for managing coastal change. By making a link between process, function and hence value of the coast, the potential for enhancing functionality through re-designing the process environment may be explored (Figure 2.3.).

Building on this development, the process–function model may be linked to an assessment of the use values associated with the range of functions provided by the coastal system. This would allow functional substitution to be examined i.e. can we maintain or increase the value of system by substituting one function for another? The most effective adaptation of such an approach into the CZM policy-making framework would be to

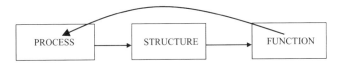

Figure 2.3: Vulnerability analysis: restructuring the basic processes to promote new or 'better' functions.

ensure that the functionality of the system becomes part of a long-term vision of the coastal zone. Such a strategy creates a long-term future for the coast that would provide for natural processes, recreation, land use and development, in a visionary but recognisable manner (e.g. long-term shoreline evolution as modelled in the UK DEFRA Future Coast project, Burgess et al., 2002).

Modelling the functions provided by the coast in the longer term would allow the processes underpinning vulnerability to be managed to either maintain or improve the functionality of the envisioned coast and provide a sustainable framework for vulnerability analysis, through which CZM policies can be developed.

Arguing the need for a comprehensive systems basis for vulnerability analysis places strong emphasis on understanding the complexities of total system response to a wide range of drivers of change: the nature of interactions between all sub-systems of the coast, and the process-based tendencies for change that are built on such relationships. Many of the chapters within this book bring important specific perspectives (whether social, economic or physical-based) on the use of vulnerability analysis: reflecting a bottom–up approach to understanding particular barriers and opportunities for CZM. This is necessary for progress towards the more effective implementation of vulnerability analysis within coastal management. However, the challenge raised by this chapter of building a series of bridges across such perspectives (with system functionality as an example) is also critical to the integrated management of the system, identifying key behavioural trends that can inform our understanding of total coastal response.

The principles presented within this chapter link the concept of vulnerability to the basic physical, social and economic processes which define the integrated behavioural of coast, embracing the entire coastal system. Regarded as a process-based potential for the system to respond to a wide range of hazard forcing, the vulnerability concept becomes a useful first-stage approach for assessing the total behavioural responses of the system given both physical and social pressures on the coast. This may be a rather ambitious approach to constructing vulnerability analysis for coastal management. However, it is only when decision making reflects such integrated dimensions of physical and cultural processes that sustainability can become a feasible goal for CZM.

Vulnerability Analysis — Wider Usefulness for CZM?

This chapter suggests that an integrated, process-based model is capable of adding significant value to the suite of tools which currently exist for CZM. Focusing on the complex behaviour of the coast, a systems approach to vulnerability analysis contributes to a fuller understanding of a primary issue of coastal management: that of the ability of society to cope with and manage coastal change.

By providing an integrative method for bringing the physical and cultural realms together within the coastal zone, vulnerability analysis enables integration of a wide range of stakeholder interests in CZM. It can bridge social and physical perspectives to provide a single conceptual framework of the coastal zone through which each of these interests can be effectively expressed. By encouraging dialogue and an understanding of the total behaviour of the coast, vulnerability analysis can contribute to furthering progress on the integrated management of coastal systems.

Vulnerability analysis can also provide an integrated and standardised framework for fragmented and incompatible information related to the nature of the coastal zone: facilitating access to effective data and information on a wide range of processes within the coastal zone. Based on an understanding of the coastal system (and not on a simple collection of environmental and socio-economic data that relate to the coastal zone) vulnerability analysis can represent an important knowledge pool of coastal behaviour.

The generally simplistic approach to the analysis also makes the complex processes which define the coast understandable to those with the responsibility for the day-to-day management of the coastal system. If the analysis is focused on the dynamics of coastal change, understanding vulnerability can help make stakeholders aware of the reality of living with the coast. It may encourage coastal managers to identify issues and management alternatives that reflect the full dimensions of system process within the costal environment, paving the way to true integrated coastal zone management.

Developing conceptual models for defining and adapting vulnerability for coastal management is critical when conducting vulnerability analysis within the coastal zone. Criteria for vulnerability assessments and methodologies for vulnerability analysis are only as effective as the conceptual framework which underpins them. This chapter highlighted the need to bring a coherent strategy for defining vulnerability within CZM: focused on a dynamic systems approach to understanding the behaviour of the coast. The chapter argues that vulnerability analysis adds value to current approaches to coastal management; that it is a flexible and adaptable concept, aiding system understanding. As such, vulnerability analysis could make an important contribution towards developing a sustainable framework for 21st century management of our coastal systems.

References

Adger, W. N. (2000). Social and ecological resilience: Are they related? *Progress in Human Geography*, 24, 347–364.

Adger, W. N., Hughes, T. P., Folke, C., Carpenter, S. R., & Rockstrom, J. (2005). Socio-ecological resilience to coastal disasters. *Science*, 309, 1036–1039.

Adger, W. N. & Vincent, K. (2005). Uncertainty in adaptive capacity. *C.R. Geoscience*, 337, 399–410.

Alexander, D. (1992). On the causes of landslides: Human activities, perception and natural processes. *Environmental Geology and Water Sciences*, 20(3), 165–179.

Bryan, B., Harvey, N., Belperio, T., & Bourman, B. (2001). Distributed process modeling for regional assessment of coastal vulnerability to sea-level rise. *Environmental Modeling and Assessment*, *6*(1), 57–65.

Burgess, K., Jay, H., & Hosking, A. (2002). *FUTURECOAST:* Predicting the future coastal evolution of England and Wales. Littoral 2002 'The Changing Coast' Porto, Portugal. *301*, 295–301.

Capobianco, M., DeVriend, H. J., Nicholls, R. J., & Stive, M. J. F. (1999). Coastal area impact and vulnerability assessment: The point of view of a morphodynamic modeller. *Journal of Coastal Research*, *15*(3), 701–716.

Cooper, J. A. G., & McLaughlin, S. (1998). Contemporary multidisciplinary approaches to coastal classification and environmental risk analysis. *Journal of Coastal Research*, *14*(2), 512–524.

de Bruijn, K., Green, C., Johnson, C., & McFadden, L. (in press). Evolving concepts in flood risk management: Searching for a common language. In: S. Begum, J. Hall, & M. J. F. Stive (Eds), *Flood risk management in Europe: Innovation in policy and practice. Advances in natural and technological hazards research*. New York: Kluwer.

de la Vega Leinert, A. C., & Nicholls, R. J. (Eds) (2001). *Proceedings of the Survas overview workshop on 'The future of vulnerability and adaptation studies'*. The Royal Chace, London, 28–30th June, 2001. London: Flood Hazard Research Centre, Middlesex University.

Downing, T. E. (2003). Lessons from famine early warning and food security for understanding adaptation to climate change: Toward a vulnerability/adaptation science? In: J. B. Smith, R. J. T. Klein & Huq, S. (Eds), *Climate change, adaptive capacity and development*. London: Imperial College Press.

Evans, E. P., Ashley, R. M., Hall, J., Penning-Rowsell, E., Saul, A., Sayers, P., Thorne, C., & Watkinson, A. (2004). *Foresight. Future Flooding Volume I – Future risks and their drivers*. London: Office of Science and Technology.

Evans, E., Hall, J., Penning-Rowsell, E., Sayers, P., Thorne, C., & Watkinson, A. (2006). Future flood risk management in the UK. *Water Management*. 159(1), 53–61.

Gornitz, V. (1990). *Vulnerability of the east coast USA to future sea level rise*. New York: NASA GSFC Institute for Space Studies and Columbia University.

Hammar Klose, E. S., Pendleton, E. A., Thieler, E. R., & Williams, S. J. (2003). *Coastal vulnerability assessment of Cape Cod National Seashore (CACO) to sea-level rise*. Open-File Report. U.S. Geological Survey.

Hammar Klose, E. S., Pendleton, E. A., Thieler, E. R., & Williams, S. J. (2004). *Coastal vulnerability assessment of Gulf Islands National Seashore (GUIS) to sea-level rise*. Open-File Report. U.S. Geological Survey.

Holling, C. S. (1973). Resilience and stability of ecological systems. *Annual Review of Ecological Systems*, *4*, 1–23.

IPCC CZMS. (1992). *Global climate change and the rising challenge of the sea*. Report of the Coastal Zone Management Subgroup Intergovernmental Panel on Climate Change Response Strategies Working Group. Intergovernmental Panel on Climate Change, United States Natural, Oceanic and Atmospheric Administration, United States Environmental Protection Agency.

Kelly, P. M., & Adger, W. N. (2000). Theory and practice in assessing vulnerability to climate change and facilitating adaptation. *Climatic Change*, *47*(4), 325–352.

Li, C. X., Fan, D. D., Deng, B., & Korotaev, V. (2004). The coasts of China and issues of sea level rise. *Journal of Coastal Research*, *43*, 36–49.

McCarthy, J. J., Osvaldo, F., Canziana, N. A., Dokken, D. J., & White, K. S. (Eds). (2001). *Climate change 2001: Impacts, adaptation and vulnerability*. Contribution of the Working Group II to the Third Assessment Report of the Intergovernmental Panel on Climate Change (IPCC). Cambridge: Cambridge University Press.

McFadden, L. (2001). *The development of an integrated basis for coastal zone management with application to the eastern coast of Northern Island*. Unpublished doctoral thesis. Queen's University, Belfast.

McFadden, L. (in press). Governing coastal spaces: The case of disappearing science in Integrated Coastal Zone Management. *Coastal Management*.

McFadden, L., Nicholls, R. J., Vafeidis, A., & Tol, R. S. J. (in press). A methodology for modelling coastal space for global assessment. *Journal of Coastal Research*.

McLean, R. F., Tsyban, A., Burkett, V., Codignotto, J. O., Forbes, D. L., Mimura, N., Beamish, R. J., & Ittekkot, V. (2001). Coastal zones and marine ecosystems. In: J. J. McCarthy, O. Canziani, N. A. Leary, D. J. Dokken, & K. S. White (Eds), *Climate change 2001: Impacts, adaptation and vulnerability*. Contribution of Working Group II to the Third Assessment Report of the Intergovernmental Panel on Climate Change (IPCC). Cambridge: Cambridge University Press.

Nicholls, R. J. (1995). Synthesis of vulnerability analysis studies. *Proceedings of the World Coast Conference 1993*. Noordwijk, The Netherlands (pp. 181–216), 1–5 November, 1993.

Nicholls, R. J., & Nimura, N. (1998). Regional issues raised by sea-level rise and their policy implications. *Climate Research, 11*, 5–18.

Nicholls, R. J., Tol, R. S. J., & Hall, J. (in press). Assessing impacts and responses to global-mean sea-level rise. In: M. Schleisinger (Ed.), *Climate impact assessment*. Cambridge: Cambridge University Press.

Pethick, J., & Crooks, S. (2000). Development of a coastal vulnerability index: A geomorphological perspective. *Environmental Conservation, 27*(4), 359–367.

Pruszak, Z., & Zawadzka, E. (2005). Vulnerability of Poland's coast to sea-level rise. *Coastal Engineering Journal, 47*(2–3), 131–155.

Sánchez-Arcilla, A., Jiménez, J. A., & Valdemoro, H. I. (1998). The Ebro delta: Morphodynamics and vulnerability. *Journal of Coastal Research, 14*(3), 754–772.

Smith, J. B., Klein, R. J. T., & Huq, S. (Eds) (2003). *Climate change, adaptive capacity and development*. London: Imperial College Press.

Turner, B. L., Kasperson, R. E., Matson, P., McCarthy, J. J., Corell, R. W., Christensen, L., Eckley, N., Kasperson, J. X., Luers, A., Martello, M. L., Polsky, C., Pulsipher, A., & Schiller, A. (2003). A framework for vulnerability analysis in sustainability science. *Proceedings of the national Academy of Sciences of the USA*, 100(14), 8074–8079.

Vogel, C., & O'Brien, K. (2004). Vulnerability and gobal environmental change: Rhetoric and reality. *AVISO*, 13, March 2004.

Yohe, G., Malone, E., Brenkert, A., Schlesinger, M., Meij, H., & Xing, X. (2006). Global distributions of vulnerability to climate change. *The Integrated Assessment Journal, 6*(3), 35–44.

Walker, B., Holling, C. S., Carpenter, S. R., & Kinzig, A. (2004). Resilience, adaptability and transformability in social–ecological systems. *Ecology and Society, 9*(2), 5. Available online-http://www.ecologyandsociety.org/vol9/iss2/art5.

Chapter 3

More or Less than Words? Vulnerability as Discourse

Colin Green and Edmund Penning-Rowsell

Introduction

The reason we seek to define vulnerability is in order to help us decide what to do to reduce that vulnerability. The value of a definition of vulnerability is consequently the degree to which it gives new and useful insights into the nature of the problem at hand. At the same time, any useful definition of vulnerability implies the adoption of a particular course of action in some specific choice, or the selection of a course of action from some set of actions where that set is specified by the definition of vulnerability.

But since different stakeholders generally have different preferences for the course of action to be adopted, and often come to the choice with strongly held beliefs as to the nature of the course of action that should be adopted, definitions of vulnerability will be strongly contested. If the stakeholders contest the choice of action to be adopted, they must also contest the definition of vulnerability. The relationship between vulnerability and the course of action is reflexive: vulnerability implies a course of action; the course of action implies a definition of vulnerability. Both also necessarily embody some claim as to the appropriate objectives that should be pursued. But the preference of one group of stakeholders for one course of action may have been formed for reasons other than a definition of vulnerability.

Defining vulnerability is thus a social act and not a technical question. It is a social act in two senses. Firstly, it is a claim that one course of action should be preferred over those courses of action preferred and perhaps proposed by the other stakeholders. Secondly, it is often a claim as to the relationships between, and roles of, the different stakeholders; it is a claim as to the basis upon which one particular definition of vulnerability should be preferred to all others.

Thus, claims by scientists to be able to give a universal definition of vulnerability are claims to access to special knowledge and understanding by reason of being a scientist. When definitions of vulnerability used by different stakeholders — including different disciplines — vary, then we are apparently faced with unresolvable claims as to the relative merit of the special knowledge and understanding of different stakeholders. Asking whether a particular definition of vulnerability is useful instead of whether it is true enables us to break out of that apparent impasse. Because vulnerability and the preferred course of action are tied together, a definition of vulnerability is also revealing of the worldview of the stakeholders who propose that definition. It is thus an aid to understanding those other stakeholders.

As a social act, a claim to a definition of vulnerability is necessarily expressed in language. If we are arguing, debating and negotiating contested definitions of vulnerability, so that vulnerability means different things to different people, then how can we communicate at all? How can we communicate when we are arguing about what we mean? Particularly when that argument is also about what should be our objectives. This is a question about the nature of language itself, as are attempts to define a unique and complete definition of vulnerability.

We therefore suggest here that attempts to reach a final and specific definition of vulnerability presuppose a particular relationship between words and concepts that would in turn render communication difficult if not impossible. Rather than an exercise in logic, we will propose that the power of language lies in its relational capacities, and particularly of metaphor and analogy.

How Can it be Useful?

If the value of any definition of vulnerability lies in the insight it can give us into the choice of the course of action to adopt, this throws the question back on to the nature of choice; what is choice and why do we have to choose?

A choice only exists if there are at least two mutually exclusive options (Green & Penning-Rowsell, 1999). So, the first condition for the existence of a choice is some form of conflict. The second condition for the existence of a choice is uncertainty (Green, 2003); if all are agreed that one course of action should be preferred over all others, then to all intents and purposes, the choice has been made. Thus, the two conditions for the existence of a choice are conflict plus uncertainty. As a result, a choice is a process through which we seek to resolve the conflicts that made the choice necessary and to become confident that one option should be preferred above all others. This process can therefore be argued to be a learning process. Two important characteristics of a choice is that about what to do; thus, it is between two or more courses of action. Secondly, it is also always prospective; any choice is an attempt to choose a future.

There are several different reasons (Green & Penning-Rowsell, 1999) why conflict occurs between the different options (Figure 3.1). Firstly, the alternatives may be functionally exclusive: for example, the choice between the use of timber groynes or fishtail rocky groynes to protect an eroding coast. Secondly, the alternatives may be mutually exclusive in time or space: someone cannot be in two different places at the same time and a wetland and a port cannot simultaneously occupy the same piece of land. Thirdly, no one option may be preferable to all others against all of the objectives we bring to that choice. As soon as the choice involves more than a single individual, then the different stakeholders can disagree as to the relative importance that should be given to achieving each of the objectives that they collectively bring to the choice.

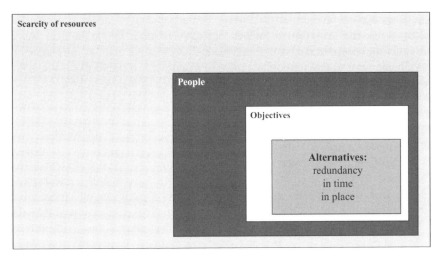

Figure 3.1: Conflict and choice.

Finally, scarcity of resources act as a constraint on choices, but in collective choices the scarcity of resources is an external constraint — a cause of conflict — on choice (Green, 2003). For example, we may all be able to agree that integrated coastal zone management strategy A is preferable to the alternative strategy B, and that transport policy M is preferable to transport policy N. But the scarcity of resources may then force us to choose between the combinations of policies A + N, and B + M. Scarcity of resources is typically an external constraint on choice rather than an internal constraint because we would still be forced to make a choice even if we had infinite resources. Thus, infinite resources would still not allow us to have a wetland and a port on the same piece of coastal land.

In collective choices, the central reason why we have to choose is that no one option is superior to all other options when compared against all the objectives that are brought to the choice. If no one option dominates all others in this way then a conflict arises when the different stakeholders disagree as to the importance that should be given achieving each of those objectives. In collective choices, one key set of objectives are those relating to the relationships between people.

When we talk of justice, equity, freedom, democracy and so on, these are all objectives that refer to how we should relate to one another. Not only are they relational, they are generally relative: this is obviously the case with equality and equity. That they are relative means that the desirability of any given action depends upon the initial starting point. If person A is initially better off than person B, then an action which results in a net gain to B and a net loss to A may be desirable. If the initial starting point is with B better off than A, then the desirability of that action may be different. Another characteristic of these relational objectives is that their basis is moral, ethical or religious: they are claims as to how we ought to relate to one another and hence as to the basis upon which choices ought to be taken.

If a choice is necessary because the objectives conflict then that choice is simultaneously a choice of means and ends. If no one option is superior to all others against all of the objectives that are brought to that choice, then any choice of means is simultaneously an assertion that the achievement of one objective is more important than any other. Hence, either the choice of means is simultaneously a choice of ends, or the choice of ends has to be made prior to any particular choice of means. It is difficult to see either why or how a choice of ends should be made prior to any choice of means, or how such a choice could be made. Hence, means and ends are the two sides of the same coin; any choice is simultaneously a choice of means and of ends. If the available means do not impact upon an objective then that objective is irrelevant to the choice. So, the introduction of a new means may invoke a new objective.

For example, until the introduction of fair trade coffee and tea, it would be difficult to introduce such objectives as equity into a choice between a cup of coffee and a cup of tea. Conversely, from the objective side of the equation, the only relevant means are those that contribute towards the achievement of one or more objectives. Hence, if it is argued that a new objective ought to be considered in making choices then it can imply that a new means should be invented that is better at satisfying that new objective than are the existing means. This has been seen with the introduction of sustainable development as an objective.

The other condition for the existence of a choice is uncertainty: doubt or uncertainty about to do, or what to choose: decision uncertainty. Such uncertainty can arise because it is not possible to resolve the conflicts that give rise to the choice; there is no clear best option.

Suppose that there were only two coastal zone management policies, A and B, and we bring only two objectives (X and Z) to the choice between them. If:

$A > B$ against objective X and
$B > A$ against objective Y and the differences are identical

then there is no rational reason for preferring A or B.

The second reason for uncertainty about what to do is the lack of knowledge; uncertainty about the world. If this is the cause, then sufficient understanding or knowledge could remove the doubt about what to do. In the above example, this occurs if we are unable to rank A and B in terms of the achievement of either or both objectives X and Y because we cannot tell how they will perform. There may, of course, be disagreements as to how any option will perform against a specific objective. The only useful information in the above example is that about the relative differences in the performance of the two options against the two objectives. Any other information would leave us better informed but not wiser about what to do.

As a process, there are many ways of making a choice; we can toss a coin, consult an oracle, or seek religious guidance. In many cultures, the preferred means of making collective choices is through the application of reason. 'Reason' is best described as a logical and rigorous process of argument, using Toulmin's (1958) definition of argument. This argument can be internal to the individual, as when the individual seeks to decide what to do, or a social process in which the stakeholders argue, debate and negotiate what course of action should be adopted.

Since choices are attempts to choose a future, it is rarely possible to demonstrate that reason has led us to make a 'better' choice in terms of selecting the best course of action than any alternative approach to making that choice. Where we are making long-term choices, we cannot look back from the future and compare the outcomes of having adopted the alternative courses of action. What reason does offer is both an intrinsic claim to superiority over the alternatives: that the use of reason is inherently desirable. As a logical and rigorous framework of analysis, which seeks to relate effect to cause, it also has two practical advantages. The rigour of reasoning creates consistency and while we may seek to be rigorous, the evidence is left to our own devices, we are not very good at it. Rigour and the resulting consistency also create accountability; it can be shown that decisions have not been taken in an arbitrary manner.

Since reason is defined as being logical, it implies a search for causation of some phenomena: why did this happen? The search for what can be done about this phenomena is the logical corollary of this question, since the choice of a course of action is logically dependent upon the causation of that phenomena. The search for causation is inherently part of reasoning even when the causes identified in a particular instance are in reality coincidental or consequent. Hence, any definition of vulnerability is also a claim to causation. That claim to causation then derives from the worldview of the individual or group making the claim; an example of such differing world views is given by cultural theory (Rayner, 1984).

Earlier we described choice as a process; as a learning process, a critical step is the invention, discovery or creation of new options. Crucially, all possible means are not known prior to the start of the process of choice; at a minimum they have to be found; they may need to be invented. In turn, vulnerability can be understood as a claim to causation; why something is susceptible to some change in its environment. That claim to causation then implies the nature of those courses of action which should be successful in reducing that vulnerability.

In summary, in seeking to derive an understanding of vulnerability — and coastal vulnerability in particular — from a theory of choice, we have suggested that:

- Choice is a process.
- Choice is necessarily simultaneously a choice of means, courses of action and of ends.
- A primary reason why collective choices are necessary is that we do not agree as to the relative importance that should be given to the achievement of different ends.
- Any claim as to the nature of vulnerability is a claim as to causation of events.
- Those claims as to causation are embedded in the worldview of those making those claims.
- In turn, definitions of vulnerability will necessarily be contested because they embody claims as to both desirable ends and of causality, and hence the adoption of some course of action naturally follows from them.
- As a process, the discovery or invention of a new option, a new course of action, is inherently part of the process of choice.
- Hence, the role of a definition of vulnerability is two-fold: to give insights into the nature of the choice which will result in the invention or discovery of a new option; and to promote understanding of the worldview of the stakeholder promoting the particular definition of vulnerability.

Vulnerability and Language

Earlier we suggested that a primary form of conflict creates choices where different individuals or groups differ in the importance they believe should be given to the achievement of different objectives, and hence which option should be chosen. In a collective choice, each of those stakeholders has to argue, negotiate and debate with the other stakeholders as what course of action should be chosen for all. They therefore use language in a way which is designed to convince the other stakeholders that their preferred option should be chosen. They seek to frame the discussion in such a way that it leads logically to the objectives that they regard as important and to the option that they favour being adopted.

Language, in short, is used purposefully with the ideal of achieving a hegemonic discourse; a frame of understanding which is adopted by all other stakeholders within the context of which discourse the choice will be made. An important instrument is the definition of vulnerability.

Hence, when Mrs Thatcher was Prime Minister, the government sought to define single mothers as moral failures and the unemployed as an underclass who were unemployed through their own failings. Defined in that way both the extent to which society as a whole had any responsibility for the conditions of single mothers and the unemployed, and the best means of delivering that responsibility, logically followed: the definition implied a very limited social responsibility towards the two groups whose vulnerability was of their own making. Conversely, if single mothers are defined as the product of gender relations, then a different social response logically follows from that definition.

Language used purposively is only one aspect of language. In discussing vulnerability, there are three key aspects of language which need to be considered:

1. the relationship between language and thought;
2. the nature of language itself;
3. how language is used.

The first, personal aspect, and the third, social aspect of language, meet in the second aspect. The personal aspect refers to the nature of thought and the relationship between language and thought. De Saussure (1983) classically argued that the mental concept (the signified) and the signifier (the material aspect e.g. the word) were inseparable (Figure 3.2). To go somewhat further, some approaches to the definition of words imply an isomorphic relationship between a word and some mental concept so that each word has a unique meaning and a single word is properly associated with each single concept.

Whorf (Carroll, 1967) went still further, arguing that language dictates what can be thought; a view now generally rejected. Wittgenstein in *Tracticus* (Quinton, 1966) took a conceptually similar approach to de Saussure, seeking to develop language as a form of symbolic logic. In essence, the attempt is to define words as atoms of meaning. If this approach is adopted then it follows that it must be possible to define a single conceptual meaning to 'vulnerability' so that using vulnerability in any other sense is both wrong and misleading. But if words bear a one to one relationship to concepts then, unless there exists some natural categories which are universally recognised (and the evidence is that there

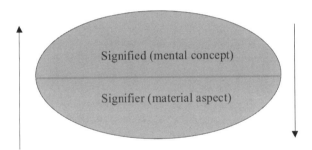

Figure 3.2: Signifier and signified.

is not (Lakoff, 1987; Rosser, 1994)), before we can use language we must first construct a socially agreed system of categorisation without using words.

But there is an alternative approach, associated both with Peirce (Hoopes, 1991) and the later Wittgenstein (Quinton, 1966). This defines words in relational terms. It is how words and concepts stand in relation to each other that give their meaning. The words 'a' and 'the' are clearly relational; there is a significant difference, for instance, between saying 'I saw a beach to-day' and 'I saw the beach to-day'. The latter implies that there is a specific beach in which the participants in the conversation share a particular interest. The words 'a' and 'the' illustrate one form of relation that can be expressed in language: drawing similarities and making distinctions, categorising the world. The most obvious relational words are referring to positions: 'above', 'left' and 'uphill' for instance. The purpose of using the term 'vulnerability' is itself relational: we are concerned to the extent to which there are differences in the degree of vulnerability.

There is always an implied 'of' and 'to' associated with the use of the term 'vulnerability', and the search for a meaning of the term is a search for a third relational term: 'because'. We seek to argue for the vulnerability *of*, say, a household *to* a coastal flood *because* of some factors. Trying to nail down the meaning of a word may therefore risk seeking to incorporate meaning into the word itself which are produced by those elements in which it stands in relation.

Metaphor and analogy are other forms of relation embodied in language. Rather than being conceptual islands referring to distinct concepts, words act as bridges. If we look at the use of commonplace words such 'head', 'table' and 'key', the different uses rely upon sharing a common characteristic with some other usage but often the same characteristic is not shared between every pair of uses (Figure 3.3). A 'table' and a 'timetable' may share a similarity of shape but, more importantly, that of organising objects in some formal order. The metaphorical power is presumably why the term 'timetable' was used rather than an entirely new term being invented for the particular purpose. It could be claimed that it is the permeability and capacity of language to change that makes it useful.

Most sweepingly, Lacan (Cobley & Jansz, 1999) argues that any single use of a word carries with it the connotations of all previous usages of the word. This offers a way of learning. This occurs, for example, in a child learning that a Persian cat and a black cat are both cats, and so in some sense is also a tiger and finally to understand what is meant if

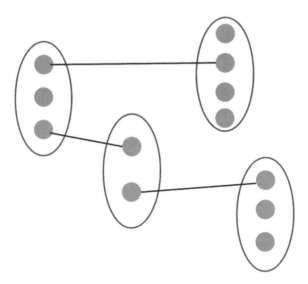

Figure 3.3: Words as metaphors.

someone is described as being 'catty'; they recall some behavioural characteristics of cats and apply those characteristics to the individual.

Alternatively, they may 'seed' a meaning to a word upon which various elaborations and interpretations of that word are created through use (Figure 3.4). The common seed meaning may be very simple and abstract; it is then the specific elaborations that developed around the seed that create something concrete and useful. For example, we use the term 'disease' to cover a multiplicity of conditions arising from multiple causes (e.g. viruses, bacteria, cysts, helminths, prions; environmental stressors such as excessive heat, excessive cold and the absence of sunlight; toxins; autoimmune problems; genetic factors; dietary deficiencies; environmental and social conditions such as those that give rise to post-traumatic stress disorder; and electromagnetic radiation).

There is little obvious similarity between, for example, smallpox, goitre, motor neurone disease, senile dementia and seasonal affective disorder — other than, as Foucault (Merquior, 1985) argues, each specifies the person in question as being unwell and as treatable by the medical professions. Thus, a particular context is associated with the stereotype. It is the particular and distinct elaboration that yields the diagnosis, the prognosis and identifies a possible course of treatment. Universal elaborations, such as possession by a devil or having been cursed, are generally held to be much less useful. However, we know what we are talking about when we speak of a 'disease'.

Thus, as metaphors, words provide useful 'stereotypes', a particular form of elaboration; 'flood' is another word where we know what we are talking about until we come to try to agree a definition. At that point, ecologists argue that floods are good things, because ecosystems depend upon both variations in flow regimes and extreme flows in particular (Acreman et al., 2000), and engineers wish to differentiate between variations in flows that do not affect human life adversely and those that do.

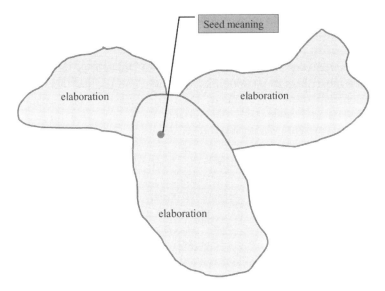

Figure 3.4: Words as seeds and elaborations.

Those stereotypes do themselves present dangers because we may then think in terms of the stereotypes which may not be either an appropriate generalisation or represent the particular individual instance. While useful for communication, the danger is that those stereotypes become taken for the whole thing; some discussions of coastal zone management then rely upon a particular stereotype or its context. Thus, it is dangerous to generalise from one country's experience to another if their contexts in terms of population density, socio-economic system and factors are different.

As the example of stereotypes illustrates, language can influence how we interpret the world. Therefore, Berger and Luckmann (1967) assert that a sociology of language cannot be separated from a sociology of knowledge; moreover, neither are neutral objective systems. By sociology of knowledge and of language they mean that they are both social constructs arising through and out of the relations between individuals and groups, and each is a reflection of the other. In turn, they contend that: *"the sociology of knowledge is concerned with the analysis of the social construction of reality"*.

We use language: we communicate with others for some purpose; one purpose is to argue with them and to convince them to adopt a specific course of action. Foucault (Merquior, 1985) has been particularly influential in stressing that knowledge is reflexive of power, that language is the operation of power. Thus, an important purpose of language is to establish a knowledge hegemony, to create power through language (Hajer, 1995).

Discourse analysis (Hajer, 1995; Dryzek, 1997) attempts to understand both the nexus of knowledge and language used by contending groups, the ways in turn that these are used in an attempt to define a particular view of reality, and persuade others that this is the appropriate understanding. Critically, discourse analysis asserts that the way in which a problem is defined determines the appropriate solutions, the ways in which frames are constructed (Fisher, 1997) imply appropriate responses. Discourse is defined by Dryzek (1997) as:

> ... a shared way of apprehending the world. Embedded in language it enables subscribers to interpret bits of information and put them together into coherent stories or accounts. Each discourse rests on assumptions, judgements and contentions that provide the basic terms for analysis, debates, agreements and disagreements. ...

Conversely, when physical scientists talk about 'risk communication' they tend to assume both that the purpose of language is to pass information and that there exists some objective reality to which those scientists have preferential access.

So, environmental groups seek to use 'natural' to imply not only good but better. Thus, for example, WWF (2002) claimed that: "Traditional forms of flood protection do not work" and that we need to "... restore our wetlands and free us of floods". WWF seeks to establish a framing in which 'natural is good and works' whereas 'traditional engineering solutions are bad and fail to work'.

Any definition of vulnerability should therefore be understood in terms of a discourse: it is a claim that the world is explicable in a particular way. The Chicago School of Geography (e.g. White, 1964) tended to frame vulnerability in terms of inappropriate behaviour on the part of those at risk, and proposed a shift away from collective action towards individual responsibility. Similarly, the Libertarian definitions of poverty (e.g. Murray, 1996) have framed it in terms of moral turpitude, of the undeserving poor who are poor because they lack character. Being vulnerable to flooding is then a sign of moral weakness.

Conversely, the social justice literature (e.g. Harvey, 1996) has argued that the poor are exposed to greater hazards because they are poor; the poor can only access the least desirable land. Similarly, in that literature it is argued that power, particularly financial and political power, is the key determinant of vulnerability. Logically, the Social Capital literature (Aldridge, Halpern, & Fitzpatrick, 2002) frames vulnerability in terms of the number and nature of relationships. The simplest model is the demographic model, which attributes vulnerability to such characteristics as age, poverty and low income. This model seems neutral but it is not to the extent to which it excludes other factors as determinants. Similarly, those models which draw only upon the properties of the world as the definers of vulnerability exclude all the other personal, social and political characteristics. What it is chosen not to include within the definition of vulnerability can therefore be as important as what is included.

A Typology of Vulnerability

If the nature of vulnerability is necessarily contested and what we hope to achieve from a definition of vulnerability is a new insight — and perhaps the discovery of a new course of action — then it follows that rather than *the* definition of vulnerability, there will be many definitions. Moreover, we also should hope for the invention of new definitions of vulnerability that give new insights. What we can do is to seek to map out the potential meanings of vulnerability.

Since vulnerability is a relational word, the first issue is to decide what is the focus of this concern: the vulnerability of what? At the coast, three systems collide: the socio-economic system, the geomorphological system and the ecosystem. Vulnerability can be defined for one, or all in conjunction. As social scientists, our personal locus of concern is on human systems, rather than the other systems, but here there are questions of and differences in scale: the individual, the household, the community or society and the associated economy as a whole. An individual household can be vulnerable to an event by which the community or society would be barely affected.

Figure 3.5 is a typology of the ways in which human vulnerability can be, and has been, defined. The first two main branches are at least superficially concerned with identifying properties, in one case of those exposed to the hazard and in the other of the hazard itself. But even here, relational issues intrude: if the characteristics of being poor and old are deemed to contribute to an individual at the coast being vulnerable (see Winchester et al., Chapter 10), is the consequence of being both additive, multiplicative or some other function? If it is not possible to define the functional relationship between the different properties considered to determine vulnerability, is anything being said at all? Moreover, measurement itself is necessarily relational; the different levels of measurement (Torgerson, 1958) that can be achieved differ both in the relationship between units of measurement (e.g. an ordinal versus an interval scale) and whether it is possible to define an absolute zero point (ratio scale) or only an arbitrary zero point (interval scale). The level at which we can measure the individual characteristics then dictates how we can relate those characteristics in the form of some measure of 'vulnerability'.

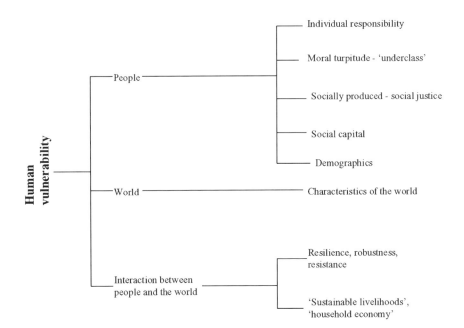

Figure 3.5: Towards to typology of definitions of vulnerability.

The third main branch in Figure 3.5 is clearly concerned with the interaction of the environment as a system and people or human systems. It is characterised by a systems approach and it is therefore concerned with dynamics. With one twig are associated those concepts of resilience, resistance and robustness which have largely been derived from ecology (Holling, 1973; Ludwig, Walker, & Holling, 1997). The other derives from development theory, notably 'sustainable livelihoods' (Ashley & Carney, 1999) and 'household economy' (Save the Children, 2000). The sustainable livelihood model takes households to have goals but to exist in an environment which is subject to unpredictable perturbations over time to which they must manage their response. The 'household economy' model was developed to explore how households cope with such a perturbation; to free up the time and resources required to cope, these must be withdrawn from other activities. Typically, this adjustment includes withdrawing girls from school and additional work for all the women in the household.

This last twig in Figure 3.5 has been argued to imply four key elements (Green, 2004):

1. It is a purposive system.
2. It is the specific objectives of the system that are important to the definition of vulnerability.
3. It exists in a dynamic environment in which the variation can help or hinder the achievement of these objectives.
4. There are a variety of potential mediating variables between the system and the environment, of adaptive strategies for the system and means of modifying the environment.

If a household is considered, then it has initially only two resources available to it (Green, 2003): time and energy (Figure 3.6). Using these it can harvest resources from the environment and convert them into household durables (e.g. dwellings, clothes) and direct consumption (e.g. food). Some forms of consumption also take time. Since humans readily form societies, an alternative mode of production is communal production, and any household has therefore to consider whether a particular form of consumption is more efficiently achieved through household production or communal production. Maintaining the social relationships that underpin that society also requires time and possibly energy as well.

If income generation is introduced into the system, the household now has four calls upon its time and energy: household production, communal production, income generation and some forms of consumption. But it can also invest some time and energy in developing multipliers which increase the productivity with which time and energy can be turned into consumption. However, the environment is not stable and the household is faced with a further choice: how much energy and time to invest in seeking to reduce the effects of perturbations. Any single perturbation must be considered in the context of other potential perturbations. In the UK, the relevant household should consider the risk of coastal flooding or erosion against such other potential perturbations as unemployment, illness and burglary. Those other perturbations may be much more important than the risk of coastal hazards.

A perturbation in the environment can then have five direct effects, which will in turn affect the household's quality of life:

1. To reduce the energy available to the household (e.g. through disease).
2. To reduce the inputs from the environment (e.g. food and water).

More or Less than Words? Vulnerability as Discourse **41**

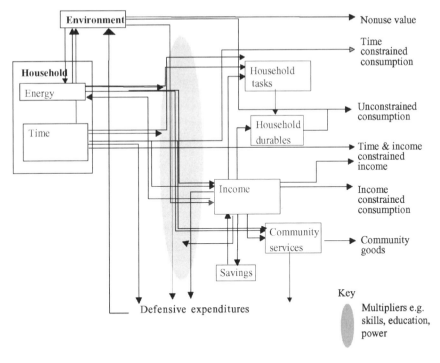

Figure 3.6: A modified sustainable livelihood model of a household.

3. To reduce income as a result of opportunities being lost or reduced to sell energy and time in the form of labour, or the returns to this time and labour.
4. Damage or destruction to household durables including the home itself.
5. A reduction in the availability of community services including health care.

In turn, there is likely to be a reallocation of time and energy from some activities to others by the different members of the household.

What insights does this analysis give us? Firstly, we cannot consider one source of perturbation in isolation. The issue is whether that particular perturbation is more significant than others. Secondly, there are three generic approaches to intervention:

1. Seeking to reduce the challenge, to reduce the magnitude of the perturbation or its nature;
2. To filter out that perturbation before it impinges upon the system in question; or
3. Reducing the impact of that perturbation upon the system by changing the nature of the system, to enhance its adaptive or coping capacity.

Assessment

In this chapter, we have suggested that, in discussing vulnerability, we need to understand both the nature of language and the purpose of the language act. To argue for a single definition of vulnerability is to argue for an isomorphic relationship between the word and a concept. Second, we claim that language is used purposefully by an individual or group in order to try to persuade others to adopt that course of action preferred by that individual or group. Hence, definitions of vulnerability are necessarily contested, not least — in the terminology of Chapter 1 — because views of adaptation and its effectiveness in reducing vulnerability are a function of views of what the ideal coastal state should comprise.

In this respect, we have suggested elsewhere (Green & Penning-Rowsell, 1999) that the coastal zone is inherently a zone of conflict, notably between those whose interest in maintaining processes (e.g. geomorphologists) and those whose primary interest is in either some current state or some other desirable state. The coastal zone is thus an area of a collision of interests as well as the collision of processes.

In turn, we will necessarily have conflicting definitions of vulnerability. Hence, the purpose of seeking to define vulnerability is to gain some new insight into what course of action we should choose. From the perspective of intervening in the interest of socio-economic systems, the interactive model of challenge (from impacts) versus adaptive coping capacity currently seems to offer the greatest insights, that is to offer the greatest scope for inventing new courses of action. But, we have to live in hope that someone will invent a new definition of vulnerability that results in new and more profound insights.

References

Acreman, M. C., Farquharson, F. A. K., McCartney, M. P., Sullivan, C., Campbell, K., Hodgson, N., Morton, J., Smith, D., Birley, M., Knott, D., Lazenby, J., Wingfield, R., & Barbier, E. (2000). *Managed flood releases from reservoirs – issues and guidance*. Report to DFID and the World Commission on Dams, R7344. Wallingford: Centre for Ecology and Hydrology.

Aldridge, S., Halpern, D., & Fitzpatrick, S. (2002). *Social capital: A discussion paper*. London: Performance and Innovation Unit.

Ashley, C., & Carney, D. (1999). *Sustainable livelihoods: Lessons from early experience*. London: Department for International Development.

Berger, P., & Luckman, T. (1967). *The social construction of reality*. Harmondsworth: Penguin.

Carroll, J. B. (1967). *Language, thought and reality: Selected writings of Benjamin Lee Whorf*. Cambridge, MA: Massachusetts Institute of Technology (MIT).

Cobley, P., & Jansz, L. (1999). *Introducing semiotics*. Cambridge: Icon.

de Saussure, F. (1983). *Course in general linguistics*. London: Duckworth.

Dryzek, J. S. (1997). *The politics of the earth, environmental discourses*. Oxford: Oxford University Press.

Fisher, K. (1997). Locating frames in the discursive universe. *Sociological Research*, 2(3). http://www.socresonline.org.uk/2/3/4.html (last accessed 2nd March 2006).

Green, C. H. (2003). *Handbook of water economics*. Chichester: Wiley.

Green, C. H. (2004). Evaluating vulnerability and resilience in flood management. *Disaster Prevention and Management*, 13(4), 323–329.

Green, C. H., & Penning-Rowsell, E. C. (1999). Inherent conflicts at the coast. *Journal of Coastal Conservation*, 5, 153–162.

Hajer, M. A. (1995). *The politics of environmental discourse, ecological modernisation and the policy process*. Oxford: Oxford University Press.

Harvey, D. (1996). *Justice, nature and the geography of difference.* Oxford: Blackwell.
Holling, C. S. (1973). Resilience and stability of ecological systems. *Annual Review of Ecological Systems, 4,* 1–23.
Hoopes, J. (1991). *Peirce on signs: Writings on semiotic.* Chapel Hill: University of North Carolina.
Lakoff, G. (1987). *Women, fire, and dangerous things.* Chicago: University of Chicago Press.
Ludwig, D., Walker, B., & Holling, C. S. (1997). Sustainability, stability and resilience. *Conservation Ecology 1*(1). http://www.consecol.org/vol1/iss1/art7 (last accessed 2nd March 2006).
Merquior, J. G. (1985). *Foucault.* London: Fontana.
Murray, C. (1996). The emerging British underclass. In: IEA (Ed.), *Charles Murray and the underclass: The developing debate.* London: IEA.
Quinton, A. M. (1966). Except from 'contemporary British philosophy'. In: G. Pitcher. (Ed.), *Wittgenstein.* London: Macmillan.
Rayner, S. (1984). Disagreeing about risk: The institutional cultures of risk management and planning for future generations. In: S. G. Haddon (Ed.), *Risk analysis, institutions and public policy.* Port Washington: Associated Faculty Press.
Rosser, R. (1994). *Cognitive development: Psychological and biological perspectives.* Boston: Allyn and Bacon.
Save the Children (2000). *The household economy approach: A resource manual for practitioners.* London: Save the Children UK.
Torgerson, W. S. (1958). *Theory and methods of scaling.* New York: Wiley.
Toulmin, S. (1958). *The uses of argument.* Cambridge: Cambridge University Press.
White, G. F. (1964). *Choice of adjustments to floods.* Research Paper No 93. Chicago: University of Chicago, Department of Geography.
WWF (2002). *Restore our wetlands or face worse floods.* Belgium: World Wildlife Fund Press Release, WWF European Policy Office.

Chapter 4

The Natural Resilience of Coastal Systems: Primary Concepts

Colin D. Woodroffe

Introduction

Coasts are particularly dynamic and the morphology of the coast is continually changing in response to various processes operating at different rates. Coastal landforms are extremely changeable and coastal habitats change over a range of spatial and temporal scales; recognition of these variations is necessary in order that planning and management can be effective.

The increasing realisation that human impacts are affecting our coastlines has promoted the concept of vulnerability. Successful management of coastlines, including mitigation of adverse impacts, must be based on an understanding of natural patterns of change. When a trajectory of change is detected, it is often difficult to determine the extent to which it is the outcome of human impact or whether it is part of the natural pattern of change that might have occurred anyway. The complexity and intricacy of the feedbacks surrounding human use of the coast and coastal resources mean that there is rarely consensus on the degree to which human actions have modified natural processes.

This chapter examines the patterns, directions and rates of change that coasts undergo. It provides a conceptual basis that underpins any consideration of the extent of human impact. The conceptual framework is illustrated with examples drawn from tropical and subtropical coasts.

Vulnerability and Resilience

Vulnerability is the degree to which a coast is likely to be affected by, or its incapability to withstand the consequences of, impact. The impact may be from a natural event, such as a storm or flood, or, as in many of the chapters that follow, it may be from human actions or events. Vulnerability to sea-level rise as a consequence of global climate change has become an issue of international concern. Impacts from other factors associated with climate change can also be anticipated, although with less certainty in terms of direction or magnitude. Vulnerability is multi-dimensional, covering natural biogeophysical response of the coast, but also involving economic, institutional and socio-cultural aspects (Klein & Nicholls, 1999). The coast can be viewed as comprising interconnected systems, a natural system and a socio-economic system.

A holistic systems approach incorporates the concept of susceptibility and sensitivity. Susceptibility describes the potential of the system to be affected, whereas sensitivity refers to its responsiveness, how likely it is to change or to fail. Its natural ability to respond can be viewed in terms of the resistance of the coast, which includes mechanical strength of materials, structural and morphological resistance, and its ability to filter the incident energy. Closely related is the concept of resilience, defined as the ability of the coast to resist change in functions or processes (McFadden, Chapter 2). In this chapter, the natural resilience of the coast is the prime focus, but it is important to recognise that similar concepts can be applied to various other aspects of the coastal management process, such as social, cultural, or institutional resilience. Resilience implies the ability of the system to bounce back, or return to some quasi-stable state. This may involve several different, though related factors, allowing a coast to withstand the failures of management, which based as it must be on incomplete understanding, is rarely ideal at protecting coastal resources.

The Coastal System

The study of the mutual co-adjustment of form and process is termed morphodynamics, and underlies our understanding of how and why landforms adjust. Coastal morphodynamic studies have led to development of physical, conceptual, mathematical and simulation models of coastal behaviour. Morphodynamic adjustments occur through the movement of sediment, and the complexity of interacting variables mean that it is useful to adopt a holistic systems approach to the coast. A system involves the interconnection of a series of variables; those within the system are dependent variables, and those outside it are called independent variables, forcing factors or boundary conditions.

Morphological States

A coastal system frequently adopts a particular 'state', defined by key parameters, of which morphology is one of the most conspicuous. Coastal landforms often show states that are in an equilibrium, or quasi-equilibrium. The system is maintained at, or more often in the vicinity of, equilibrium by several negative feedbacks and may change between states as a result of changes in boundary conditions. Coastal systems are generally complex non-linear dynamic systems. Equilibria are recognised by persistence of some morphological feature. For example, a beach is an accumulation of loose sand, every grain of which can be moved by the wave energy to which the beach is subject periodically, if not continually. It undergoes changes in shape as an outcome of entrainment and re-deposition of sand, moulding the beachface and associated surf zone into a particular state to either reflect or dissipate the energy of the waves. A beach persists as a result of the tendency for self-organisation through complex feedbacks (Short, 1999).

There has been considerable debate about the extent to which beach shape represents an equilibrium in profile, and in planform. An equilibrium is called 'regime' in the engineering literature, and empirically calibrated rules for the offshore shape of the shallow nearshore profile, and log-spiral shapes for beaches in planform have been proposed. Such an equilibrium is considered an attractor in the language of chaos. It has been described as representative of ecosystem 'health' in terms of the ecology of the system. When monitored over time the shape of a beach is usually found to change subtly, although through a limited number of forms, termed beach states (Figure 4.1).

Beaches adopt one of several states, and sophisticated models have been developed that link beach states with incident wave energy. Initially these models were developed for wave-dominated beaches along the coast of southeastern Australia, incorporating the formation of nearshore bars, changes in beach slope or other parameters. More recently it has been shown that the broad continuum from reflective to dissipative beach states can be applied to beaches around the world (Short, 1999). Beaches undergo change between beach states in response to variations in external factors, such as climate and wave conditions, termed boundary conditions. Figure 4.1 shows schematically the morphological change of a simple wave-dominated beach. The beach in planform lies between two headlands. Its profile can be seen to vary from an accreted form with a pronounced beach berm and a steep beach face (termed reflective because wave energy is predominantly reflected back off the beach face) to an eroded form in which the beach is flatter and a considerable volume of sand has been removed from the beach and is stored in the nearshore in the form of a bar

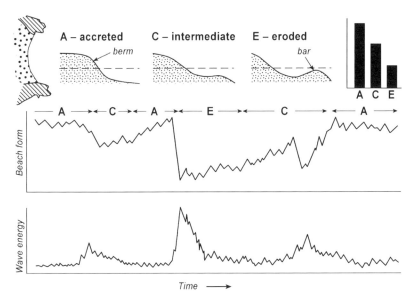

Figure 4.1: Beach morphodynamic concepts illustrated for a schematic beach between headlands. The beach may adopt accreted (reflective), intermediate or eroded (dissipative) states. Its shape, measured by some parameter such as subaerial sand volume, varies in response to wave energy. The response of beaches to wave energy is not immediate, after a storm event during which the beach adjusts to its eroded state there is a gradual recovery.

(termed dissipative because wave energy is dissipated in the surf zone as waves break over the bar). A major storm has the effect of eroding the beach, which adjusts its form to become flatter and builds a nearshore bar that dissipates wave energy. Erosion of the beach occurs during a storm, but that form may persist, and it usually takes weeks or months for the beach to build back to its accreted state. Over time, some beaches that are subject to variable wave energy, fluctuate in response to incident wave energy; as shown schematically in Figure 4.1, and the state which occurs for most of the time, is termed the modal state. More detailed descriptions of these beach morphodynamic models are available in Short (1999).

Types of Equilibrium

If the boundary conditions that affect a beach remained constant then the equilibrium shape of that beach would not be expected to change. It is more often the case for geomorphological systems that external conditions do change and there is a dynamic equilibrium. This is especially true of coastal systems in which sediment is either being deposited or being eroded. Figure 4.2 shows an example of three different beaches each of which adopts a different sort of equilibrium. A sheltered beach may be so immune to changes in wave energy that it remains in a static equilibrium. A more exposed beach may respond, as illustrated in the example in Figure 4.1, by adjusting between an accreted and eroded state, and is termed a metastable equilibrium; one state is found under regular conditions, and a second occurs under a higher energy impetus. Where the volume of sediment on a beach is gradually increasing, as for example where a river supplies sediment to the beach compartment, the beach adjusts in dynamic equilibrium (Woodroffe, 2003).

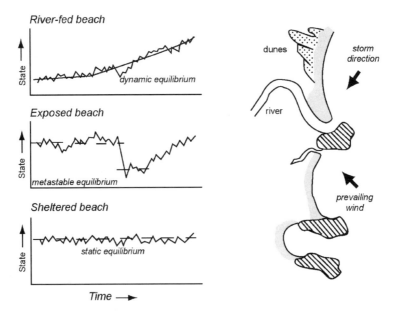

Figure 4.2: Three beaches adopting a different type of equilibrium (after Woodroffe, 2003), see text for details.

The concept of equilibrium is represented by analogy in Figure 4.3, in terms of balls and their movement across a landscape which has a series of depressions. Such an analogy has been widely used in systems literature (e.g. Scheffer, Carpenter, Foley, Folke, & Walker, 2001), where the ball can be thought of as in equilibrium when it is at rest. In a coastal context each ball could be thought of as a pebble on the shore. Simple equilibria can comprise several ideal situations. A stable equilibrium is one in which there appears to be no change and processes are balanced (an attractor); any slight disturbance may move the ball, but it returns, or is attracted, to the bottom of the depression. This is in contrast to an unstable equilibrium (repellor). Unstable equilibria can exist, but the slightest perturbation is likely to disturb the balance and the system then accelerates away from that state; in the case in Figure 4.3 the ball may settle on the top of the crest, but the slightest impulse will result in it rolling away.

Static equilibrium is where no change occurs, and is defined by persistence of the state. Steady-state equilibrium is where boundary conditions do not alter so the system demonstrates stationarity. Some systems adopt a metastable equilibrium, they can occupy two states, but require additional energy to move to their higher energy state; the eroded beach in Figures 4.1 and 4.2 is an example, whereas in the case of pebbles on the shore, it may take a larger wave (energy input) to raise a pebble to a hollow higher on the shore. Dynamic equilibrium is a complex and confusing concept, but it can be thought of as the sort of balance that persists where the shape of the landscape itself is evolving, a situation that is common in geomorphology. The ball in this case is striving to adjust to a moving target (Figure 4.3).

Differences in the Response of Coastal Systems

Any framework within which resilience or vulnerability of the coast is examined, needs to consider several of the key factors influencing the way the physical systems on the coast

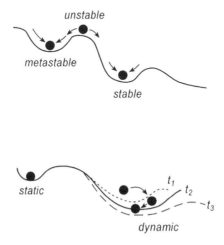

Figure 4.3: The concept of equilibrium. Upper diagram shows stable, unstable and metastable equilibria, lower diagram shows static and dynamic equilibrium.

behave. Adjustments of coastal systems can occur at any of a series of spatial and temporal scales; they occur at varying response times, cross critical thresholds beyond which behaviour changes, and are influenced, in often uncertain ways, by previous sets of conditions. These factors are examined in this section.

Temporal and Spatial Scales

Coastal morphodynamics operate within a hierarchy of temporal and spatial scales (Cowell & Thom, 1994). These are shown schematically in Figure 4.4 with examples from reef systems as an illustration. Time is generally treated as linear and progressive, but in some instances it can be circular, or cyclic. The smallest scale is the 'instantaneous' scale. This is the time frame within which individual waves occur where the physics of fluid dynamics apply (Figure 4.4) and is best examined at a very local spatial scale. At this scale, the linear equations of physics apply and can be expected to have predictive value to the extent that these processes are understood or can be measured. In the case of a reef, the instantaneous scale covers the physical processes beneath an individual wave and the biological processes which enable the coral to grow and so produce sediment (which is the product of the breakdown of the skeletons of coral and other carbonate organisms living on the reef, such as coralline algae, molluscs and foraminifera).

A longer time scale is the 'event' scale, at which a perturbation occurs and the system responds. In the case of reefs it is processes at the event scale, such as tropical storms,

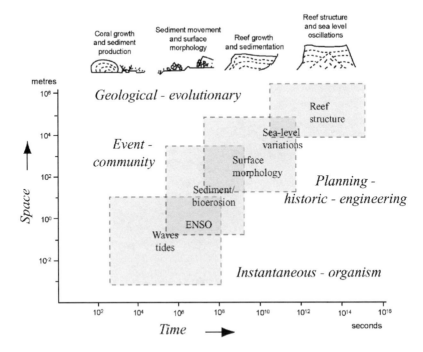

Figure 4.4: Temporal and spatial scales (based on Cowell & Thom, 1994, modified from Woodroffe, 2003), and significant processes on reefs (see text for details).

which have the greatest control on the surface morphology of reefs, detaching and breaking coral colonies and building rubble ridges. The effects of a period of high wave energy have a similar impact on reefs to that shown on beach erosion in Figure 4.1. In Figure 4.1, the beach responds to the storm (high wave energy) by changing state and then recovers from this event towards its pre-storm state (a measure of its resilience). This event scale is described as 'short-term' by Brunsden (2002), and in geomorphology he regards that predictive inferences at this scale are at best informed guesses.

The next level of study concerns regional scales over decades to centuries. In the case of reefs (Figure 4.4), changes at this scale are preserved in the pattern of reef growth and sedimentation. Termed long-term by Brunsden, this is a time scale of particular significance in the context of human societies. It has been variously referred to as the 'historic', 'engineering' or 'planning' time scale (Woodroffe, 2003) because it is the scale over which we know from historical records that there have been changes, and which is of especial significance in terms of planning or engineering projects. Brunsden considers that any attempt to forecast at these scales can be thought of as prophecies, the coastal system's broad behavioural patterns may be known but cannot be predicted with any certainty.

The largest scale is the geological time scale and the global spatial scale. This is the time scale of millennia, extending to millions of years. At this scale there are important broad trends with very significant implications, for example Quaternary variations in sea level have seen significant movements of the shoreline both vertically and horizontally. In the case of reefs, there are remnants of former reefs that grew during previous interglacials and which give valuable insights into tectonic movements, whether uplift or subsidence, and which have constrained our understanding of global climate change. However, the processes at work at these time scales may be imperceptible in the day-to-day management of coastal systems.

Response Time

Time scales become important in the context of forcing functions, or perturbations to the system, termed stresses, or stressors in ecological literature. These are triggering events, representing an energy input to which the system may respond. It is important to recognise several parameters relating to these events, they vary in intensity, duration and frequency; they may be acute, episodic or periodic. Response time to events comprises reaction time, the time it takes for the system to react, and relaxation time, the time it takes for the system to regain its pre-disturbance condition. The average time between events of each magnitude is called the recurrence interval and has a particular significance in terms of whether or not the system has time to recover before there is another event. The reccurrence of events, called event sequencing, can be significant; where one event is followed by another before it has had time to recover, the effect of the second event may then be very different from that of the first. Reaction time, and more particularly relaxation time, may be delayed (lagged). If the system does return to its previous condition then events represent perturbations or pulses, and the system is called intransitive or pulsed (Brunsden & Thornes, 1979; Chappell, 1983). If the system is not stationary in time, and does not recover before the next event, then it is called ramped or transitive, and in this case its post-event condition is different from its pre-event condition.

Time is also important because different things occur at different time scales. Geomorphological changes are usually slow, especially in comparison with most of the time scales of human enquiry, for example monitoring has rarely been carried out for more than a few decades, and 'thesis' time over which intensive studies are carried out is generally only a few years. Other time frames, such as political life (one election term, or one term of office?), human generations, and design time for engineering projects are also short in comparison with the time it has taken for coastal landforms to evolve to their present state.

Incompatibility of time scales can be seen where different disciplines try to come together; for example hydrodynamic adjustments occur in instantaneous time, whereas morphological adjustments, requiring time for the movement of sediment, occur over slower time scales. Many hydrodynamic models are essentially static, they model conditions at the time of observation; dynamic models, which adjust the morphology based on modelling outcomes for process operation, are generally complex and constrained in terms of the time steps they use (see for example van Rijn, 1993).

Models are simplifications that capture the essential behaviour of systems, but they are generalisations and are only indicative, many are probabilistic. Until recently there has been a lack of models for complex non-linear processes, which has hindered detailed quantitative impact assessment. Considerable effort is being made to develop models of large-scale coastal behaviour, attempting to scale up short-term modelling to have greater relevance at longer time scales, but such models are still in their early stages (de Vriend et al., 1993).

Thresholds and Self-Organisation

Thresholds are important steps in system development. A system may pass a threshold in response to external boundary conditions (such as sea level), intrinsic triggers, or a perturbation to the system. Negative feedback (self-regulation) tends to keep a system in, or searching around, an apparent equilibrium. Many systems show positive feedback, or self-organisation, whereby a system may develop along a trajectory of accumulated geomorphological change. Such a trajectory can cross an intrinsic threshold, an abrupt change that occurs without external stimuli.

Figure 4.5 shows a schematic pattern of change on an intermittently open coastal lagoon or barrier estuary. Systems of this type, which are sometimes connected to the sea, but at other times separated from the sea by a sand barrier, are characteristic of the smaller estuaries along the coast of southeastern Australia and in southern Africa. When river discharge is sufficient there is an inlet through which tidal exchange occurs. The tidal processes build a tidal delta with sand accumulating both as an ebb delta on the oceanward side, but more particularly as a flood tide delta on the landward side. Persistent swell activity continues to supply sand to the inlet and during periods of low terrestrial inflow the inlet may close, sealing the coastal lagoon off from the ocean. In the example in Figure 4.5 it is suggested that once the inlet is sealed it requires a higher water level, triggered by a large rainfall event in the catchment, to cross the threshold and reopen the inlet.

There is a tendency for systems to become more organised through time. If a large pile of sand is dumped on a beachface, wave energy is likely to redistribute that sand and to adjust towards an equilibrium beach as described above. The process can be observed as

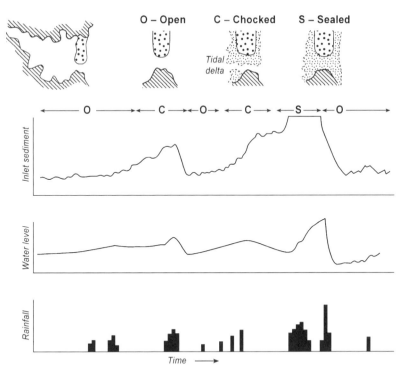

Figure 4.5: Coastal lagoon or barrier estuary, as found along coast of southeastern Australia, showing three states of the inlet or entrance, and a schematic illustration of how that entrance responds to rainfall events and subsequent water level.

the waves rework a sand castle built when the tide is low, and similar processes reshape a beach that has been nourished with sand, as occurs in beach replenishment schemes. If a large pile of very mixed sediment sizes is dumped on the beachface, the various grain sizes become progressively sorted; fine sediment is washed away, sand of similar size to the beach incorporated into the beach, and large boulders left as lag. These processes of organisation can be seen to operate where glacial deposits of highly mixed sediments are reworked by wave action. A striking example is the detailed comparison of sequences of eroding drumlins along the coast of Nova Scotia; mixed moraine from drumlins becomes sorted as the different grain sizes are moved differentially from the cliffed face of the drumlin. Swash-aligned coarse boulder banks develop in front of the cliff face, finer material becomes incorporated into drift-aligned spits, and a self-organised sequence of coastal landforms evolves (Orford, Carter, & Jennings, 1991).

Inheritance and State-Dependence

An important difference between the timeless experiments of physicists or chemists and the landform systems of the geological scientist is that the geological evolution of landforms is time-bound. The state that develops is contingent on events that have gone before. This is generally termed inheritance, or state-dependence. The term sensitive-dependence on initial conditions has become widely used in the language of chaos theory, but this pre-supposes the time-bound laboratory environment in which initial conditions can be defined. Initial conditions cannot be known for the ongoing experiments that nature runs in the real world, increasing the uncertainty about how a coast will evolve (Phillips, 1996). Coastal landforms are often partly contingent on previous landforms or sets of conditions. As a consequence, the evolution of a set of coastal landforms is unpredictable, unrepeatable and irreversible (Cowell & Thom, 1994). The present state is partly an outcome of unique past events; beach states, for example, are not solely a function of contemporaneous wave conditions, but inherit a form from previous wave conditions and beach states. Although there is considerable uncertainty about the details, the broad domain within which the coast operates can be known, providing a range within which probabilistic models can be developed, rather than deterministic models.

In fact, states are re-adopted, but the pathway is often not quite the same in both directions. For example, beaches erode and subsequently recover through a series of intermediate beach states (Figure 4.1). Where the pathway from one state to another takes one route but returns by another, this is termed hysteresis. Formative events are not necessarily extreme events. The critical threshold at which landform change occurs can change over time; for example in a cliff cut into two lithologies, small collapses of the lower lithology may cause the lower strata to retreat to the point where failure in the upper strata occurs, though not in response to one particular extreme event, rather as the outcome of the lower face passing a critical threshold (Brunsden, 2001).

In practice, the coast comprises many complex systems which interact and which may be coupled, as in the cliff collapse example. The pattern of opening and closing of coastal lagoons along the southeast Australian coast and the southern coast of South Africa is more complex than shown in Figure 4.5 and is a function of both conditions to landward in the lagoon, particularly water level, and the state of the beach on the sand barrier (in particular the beach state as indicated in Figure 4.1). Figure 4.6 attempts to capture how these two forcing functions might be coupled. The schematic representation suggests that water level alone may not be sufficient to breach a sealed barrier and reopen an inlet. The behaviour

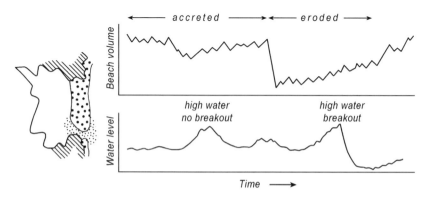

Figure 4.6: Coastal lagoon showing coupling between beach states (as shown in Figure 4.1) and water level (as shown in Figure 4.5).

of the beach is also important; if the beach is accreted then it is less likely to be breached at a critical water level, whereas if it is eroded, the threshold water level at which reopening occurs may be lower. Armouring of a beach represents another example of a time-dependent process that lessens the likelihood of change and alters the threshold at which other adjustments to landforms can occur (Brunsden, 2001).

Equilibrium and the Resilience of Coastal Systems

The existence of conditions that recur or persist, variously termed 'states', 'equilibria' or 'attractors', each having some stability as a result of negative feedback suggests that coastal systems are resilient (Gunderson & Pritchard, 2002). It is possible to identify some subtle differences in how different disciplines have defined a system's resilience and a range of definitions are illustrated schematically in Figure 4.7. Gunderson, Holling, Pritchard, and Peterson (2002) discriminate an engineering and ecological definition, to which is added a geomorphological definition (Brunsden, 2001). The engineering definition involves a measure of the time it takes to return to equilibrium; in Figure 4.7 the steep sided 'attractor' implies a rapid return to equilibrium (resilience). This definition assesses the resistance of the system; an engineering solution such as a seawall is built to resist change. The ecological definition of resilience follows an approach advocated by Holling (1973) who considered it a measure of the ability of a system to absorb changes. It is sometimes measured as the speed of return to the original state (Pimm, 1984). In this case, equilibria are thought

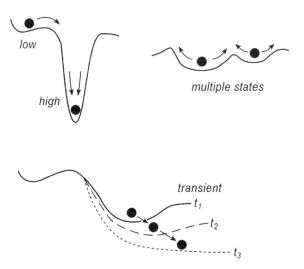

Figure 4.7: Schematic representation of different concepts of resilience. Engineering resilience is shown where structures are designed to be resistant; ecological resilience is where there is elasticity of ecosystems and may involve multiple states (based on Gunderson et al., 2002), and geomorphological resilience involves dynamic systems where the equilibrium may change over time.

of as broad states and resilience involves the breadth of the range over which a system may return to its previous equilibrium state as opposed to adopting an alternative state. Expressed differently, it is the magnitude of disturbance that can be absorbed before the system redefines its structure. Ecological resilience relies on diversity of species, and particularly suites of species that fulfil functional roles; it may deteriorate over time after repeated disturbance (Gunderson et al., 2002).

Coastal and marine ecosystems are often considered in terms of multiple states, and it has been argued that more than one stable community of organisms may be possible in a given habitat (Knowlton, 2004). A change of state, or phase shift, appears to have occurred on coral reefs, suggesting that coral cover, or algal cover may represent alternative stable states (Hughes & Connell, 1999). The significance of this concept for coral reefs is examined below.

Brunsden (2001) has defined geomorphological resilience as the degree to which a system recovers to its initial pre-disturbance state. This recognises that many landform systems strive to reach a dynamic equilibrium, and that the system may not return to an identical state after a disturbance. Brunsden identifies that change between alternative states may be prevented by some barrier to change (the lip of the depression in Figure 4.7), but that this may be transient. Concepts of the elasticity of the system and its malleability are closely linked with this view of resilience. Instead of the static view that is necessary when the engineer is considering the design life of a structure, geomorphologists recognise that landform systems are rarely stationary in time, as boundary conditions change at various time scales. In the case of the coast, sea level is one boundary condition which is known to have changed, and can be anticipated to change in the future. Sea level adjusts at several time scales, and has a profound impact on the coast. This is examined in the coral reef examples discussed below.

Resilience is considered to encompass different things by different researchers, and has been extended to include socio-economic systems. Klein, Nicholls, and Thomalla (2002) have suggested that it has become largely meaningless as a term unless the sense in which it is used is defined. They choose to define resilience in terms of two system attributes, the amount of disturbance a system can absorb and still remain within the same state, and the degree to which it is capable of self-organisation to preserve its actual and potential functions. The term adaptive capacity is widely advocated, as outlined elsewhere in this book, to cover how human adjustments may be incorporated along with the natural variability of the system.

Coral Reefs and the Resilience of Reef Systems

In the case of coral reefs, growth of corals and associated organisms directly influences the morphology of the reef, and reefs can be viewed both as ecological and as geomorphological systems. A coral reef is an accumulation of carbonate, dominated by coral framework, but also incorporating bioclastic sediments derived from other calcareous organisms (coralline algae, molluscs and foraminifera). Coral reefs occur where environmental conditions, such as wave energy, water temperature and water depth are favourable, and the distribution of different growth forms of coral is linked to environmental gradients in light availability, wave energy and sedimentation. Reefs are limited to the photic zone by symbiotic zooxanthellae. In addition to wave energy and nutrient availability, they are constrained in their upper growth by subaerial exposure at lowest tide levels. Corals may be replaced in extremely high wave-energy settings, or at suboptimal water temperatures, by coralline algae. Reefs offer an unparalleled opportunity to examine the nature of past coasts over the full range of time scales identified above as a result of the relatively good preservation of reef limestone and associated sediments (shown schematically in Figure 4.4).

The response of reefs to sea-level change provides an example of the way that a coastal system adjusts to a boundary condition (Figure 4.8), and one that may yield insights into the likely impact of future changes in sea level on reefs. Although individual corals can grow at rates of 10–100 mm a^{-1}, the consolidation of reefal material into a reef occurs more slowly and reef accretion rates vary in the range of 1–10 mm a^{-1}. Where rates of sea-level rise are very rapid, the reef is drowned. At slightly slower rates of rise the reef is likely to backstep, as appears to have occurred in the West Indies around 7 ka BP (Neumann & Macintyre, 1985). If rates of sea-level rise decelerate, the reef has the opportunity to catch up with sea level, as has occurred on much of the Great Barrier Reef (Davies & Montaggioni, 1985). Where the rate of rise is similar or less than the rate of reef growth, a reef can keep up with sea level, as has been the case on the barrier reef in Tahiti (Montaggioni et al., 1997). If sea level is stable, once a reef has reached sea level it will prograde, although several different modes of reef progradation are possible (Kennedy & Woodroffe, 2002). If sea level falls, the reef that has grown up to a higher sea level is left as an emergent reef flat, as is common on many reefs in the Indo-Pacific, explaining the broad reef flat, largely bare of live coral, found on many mid-Pacific atolls (Woodroffe, 2003). Each of these sea-level and reef-growth scenarios can be illustrated

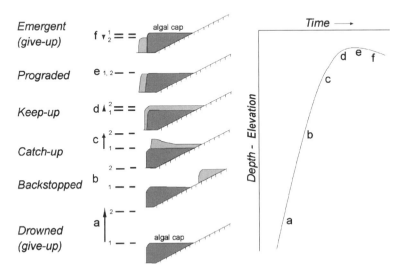

Figure 4.8: Response of reef stratigraphy to sea-level change (after Woodroffe, 2003), see text for details.

by different reefs from around the world, illustrating the range of responses that reef morphology can show to variations in the rate of sea-level change.

In contrast to the view that tropical ecosystems are diverse because they have remained unchanged for millions of years, it is now widely accepted that coral reefs are subject to frequent disturbance. Reefs appear to have adapted to episodic perturbations and demonstrate ecosystem resilience (Brown, 1997). However, it is important to recognise that physical processes and biological processes operate over different time scales (Hatcher, 1997). Thus, while plate tectonics, evolution and mass extinction occur over millions of years, sea-level fluctuations and reef growth are more apparent over millennia (see Figure 4.4). On the other hand, decline of coral reefs through historical time, as a result of human over-exploitation, particularly overfishing, has been exacerbated as a consequence of short-term perturbations, such as individual storms, epidemic disease and El Nino-Southern Oscillation events that recur at shorter time scales than the life history of many of the more massive coral species.

Many reefs appear to be a temporal mosaic of communities at various stages of recovery from these various short-term disturbances (McManus & Polsenberg, 2004). However, the resilience of coral reefs has been called into question in view of an apparent phase shift from coral-dominated reef to one that is dominated by algae. It appears that several types of algal-dominated community (calcareous encrusting algae, low-stature turf algae, calcareous frondose algae, fleshy macroalgae) can exist as alternative states on reefs, but there is concern as to whether coral cover will re-establish, either through regrowth of existing colonies (resheeting) or new recruitment. This has been particularly expressed in relation to Caribbean reefs where a series of separate disturbances, including damage by hurricanes, bleaching as a result of thermal stress, disease and eutrophication, appear to have reduced the capacity of coral to recover (Lesser, 2004). Gradual, but constant, stresses, particularly those resulting from human impacts, may push reefs beyond a resilience threshold.

Reefs appear resilient in view of the robustness shown by their persistence in the geological record. They appear to have coped with major changes of sea level that have completely displaced entire reef structures vertically and horizontally. By contrast with this robustness, reefs also appear fragile and sensitive to changes in environmental conditions (Done, 1999). Widespread coral bleaching, detected on an unprecedented scale around the globe in response to El Niño-related warming in 1998, poses a particular threat to reefs. Bleaching occurs when warmer that usual sea-surface temperatures lead to expulsion of the symbiotic zooxanthellae, and the coral surface becomes pale, in many cases leading to mortality. Global warming, as a result of the enhanced greenhouse effect, poses a particular threat to reefs and there is a vigorous debate over whether reefs are resilient enough to be able to survive (Douglas, 2003). The synergistic effects of various others pressures, particularly human impacts such as overfishing, appear to be exacerbating the stresses on reef systems and, at least on a local scale, exceeding the thresholds beyond which coral is replaced by other organisms.

Summary

A change in the state of a coastal system can occur as a result of one of three factors: (a) a short-term response to a perturbation, (b) as a result of the system passing an intrinsic threshold, or (c) in response to a change in boundary conditions. It is important to understand which of these causes has led to any observed change in natural coastal systems. Coastal systems undergo changes on a range of scales, some of which may be short term, others of which may be lagged responses or intrinsic thresholds. These have been illustrated with examples of beach and coastal lagoon behaviour. Coral reefs leave an incomplete record of past changes within the limestone of the reef, interpretation of which may provide clues to the way that reefal environments have adjusted in the past. Human factors can be, and very often are, associated with changes to the coast. It is important to be able to discriminate between natural adjustment and adjustments that have been exacerbated by human action. It will be crucial to discriminate between a reef's ability to withstand a gradual change in a boundary condition and a more rapid human-induced alteration. Not only will this require a good understanding of how the natural system adjusts, but it will also require scientists to adopt more rigorous adjudication. There are likely to be a wide range of social, cultural and political reasons that obscure the role that humans have had, or are having, in changing the way that coastal systems operate.

Acknowledgements

I thank Lorraine McFadden for the invitation to contribute to this book and Piet Huizinga for helpful comments on the behaviour of intermittently open coastal lagoons on the coast of South Africa.

References

Brown, B. E. (1997). Adaptations of reef corals to physical environmental stress. *Advances in Marine Biology, 31*, 221–299.

Brunsden, D. (2001). A critical assessment of the sensitivity concept in geomorphology. *Catena, 42*, 99–123.

Brunsden, D. (2002). Geomorphological roulette for engineers and planners: Some insights into an old game. *Quarterly Journal of Engineering Geology and Hydrogeology, 35*, 101–142.

Brunsden, D., & Thornes, J. B. (1979). Landscape sensitivity and change. *Transactions of the Institute of British Geographers, 4*, 463–484.

Chappell, J. (1983). Thresholds and lags in geomorphologic changes. *Australian Geographer, 15*, 357–366.

Cowell, P. J., & Thom, B. G. (1994). Morphodynamics of coastal evolution. In: R. W. G. Carter, & C. D. Woodroffe, (Eds), *Coastal evolution, late quaternary shoreline morphodynamics*. Cambridge: Cambridge University Press.

Davies, P. J., & Montaggioni, L. F. (1985). Reef growth and sea-level change: The environmental signature. *Proceedings of the 5th International Coral Reef Congress, 3*, 477–515.

de Vriend, H. J., Capobianco, M., Chesher, T., de Swart, H. E., Latteux, B., & Stive, M. J. F. (1993). Approaches to long-term modelling of coastal morphology: A review. *Coastal Engineering, 21*, 225–269.

Done, T. J. (1999). Coral community adaptability to environmental change at the scales of regions, reefs and reef zones. *American Zoologist, 39*, 66–79.

Douglas, A. E. (2003). Coral bleaching — how and why? *Marine Pollution Bulletin, 46*, 385–392.
Gunderson, L. H., Holling, C. S., Pritchard, L., & Peterson, G. D. (2002). Resilience of large-scale resource systems. In: L. H. Gunderson, & L. Pritchard (Eds), *Resilience and the behaviour of large-scale systems*. Washington: Island Press.
Gunderson, L. H., & Pritchard, L. (2002). *Resilience and the behaviour of large-scale systems*. Washington: Island Press.
Hatcher, B. G. (1997). Coral reef ecosystems: How much greater is the whole than the sum of the parts? *Coral Reefs, 16*, S77–S91.
Holling, C. S. (1973). Resilience and stability of ecological systems. *Annual Review of Ecological Systems, 4*, 1–23.
Hughes, T. P., & Connell, J. H. (1999). Multiple stressors on coral reefs: A long-term perspective. *Limnology and Oceanography, 44*, 932–940.
Kennedy, D. M., & Woodroffe, C. D. (2002). Fringing reef growth and morphology: A review. *Earth Science Reviews, 57*, 257–279.
Klein, R. J. T., & Nicholls, R. J. (1999). Assessment of coastal vulnerability to climate change. *Ambio, 28*, 2, 182–187.
Klein, R. J. T., Nicholls, R. J., & Thomalla, F. (2002). *The resilience of coastal megacities to weather-related hazards; a review*. Workshop: The future of disaster risk: Building safer cities. 4–6 December 2002, Disaster Management Facility, World Bank.
Knowlton, N. (2004). Multiple "stable" states and the conservation of marine ecosystems. *Progress in Oceanography, 60*, 387–396.
Lesser, M. P. (2004). Experimental biology of coral reef ecosystems. *Journal of Experimental Marine Biology and Ecology, 300*, 217–252.
McManus, J. W., & Polsenberg, J. F. (2004). Coral–algal phase shifts on coral reefs: Ecological and environmental aspects. *Progress in Oceanography, 60*, 263–279.
Montaggioni, L. F., Cabioch, G., Camoinau, G. F., Bard, E., Ribaud Laurenti, A., Faure, G., Dejardin, P., & Recy, J. (1997). Continuous record of reef growth over the past 14 k.y. on the mid-Pacific island of Tahiti. *Geology, 14*(25), 555–558.
Neumann, A. C., & Macintyre, I. (1985). Reef response to sea level rise: Keep-up, catch-up or give-up. *Proceedings of the 5th International Coral Reef Congress, 3*, 105–110.
Orford, J. D., Carter, R. W. G., & Jennings, S. C. (1991). Coarse clastic barrier environments: Evolution and implications for Quaternary sea level interpretations *Quaternary International, 9*, 87–104.
Phillips, J. D. (1996). Deterministic complexity, explanation, and predictability in geomorphic systems. In: B. L. Rhoadds, & C. E. Thorn (Eds), *The scientific nature of geomorphology*. Chichester: Wiley.
Pimm, S. L. (1984). The complexity and stability of ecosystems. *Nature, 307*, 321–326.
Scheffer, M., Carpenter, C., Foley, J. A., Folke, C., & Walker, B. (2001). Catastrophic shifts in ecosystems. *Nature, 413*, 591–596.
Short, A. D. (1999). *Handbook of beach and shoreface morphodynamics*. Chichester: Wiley.
van Rijn, L. C. (1993). *Principles of sediment transport in rivers, estuaries and coastal seas*. Amsterdam: Aqua Publications.
Woodroffe, C. D. (2003). *Coasts: Form, process and evolution*. Cambridge: Cambridge University Press.

Chapter 5

Integrating Knowledge for Assessing Coastal Vulnerability to Climate Change

Jochen Hinkel and Richard J.T. Klein

Introduction

In view of the high natural and socio-economic values that are threatened and might be lost in coastal zones, it can be important to identify the types and magnitude of changes to which coastal systems are exposed, as well as the options that are available to minimise risks and reduce possible adverse consequences. However, assessing coastal vulnerability is not a straightforward exercise, not in the least because there is confusion concerning the precise meaning of the term "vulnerability". Vulnerability is specific to a given location or group or sector. There is therefore no single recipe for assessing vulnerability to climate change or any other type of change. Different scholarly communities have developed different conceptualisations of vulnerability, and different conceptualisations exist even within these communities.

Existing conceptualisations are often found to be imprecise when attempting to make them operational for assessment. For example, the Third Assessment Report of the Intergovernmental Panel on Climate Change (IPCC) defined vulnerability as "the degree to which a system is susceptible to, or unable to cope with, adverse effects of climate change, including climate variability and extremes. It is a function of the character, magnitude and rate of climate variation to which a system is exposed, its sensitivity, and its adaptive capacity" (McCarthy, Canziani, Leary, Dokken, & White, 2001). The extent to which this "definition" of vulnerability can be made operational is limited because its constituent concepts are either very general or remain undefined. In addition, it does not explain the functional relationships between these concepts.

The diversity in conceptualisations and their being imprecise have led to a diversity in methodological approaches for assessing vulnerability (Brooks, 2003; O'Brien, Eriksen, Schjolden, & Nygaard, 2004; Adger, 2006). In addition to this methodological diversity, methodologies have grown in complexity over the past two decades: they now consider multiple stimuli rather than a single stimulus, they allow for dynamic rather than static analysis, they have become interdisciplinary and they have moved from a predominant emphasis on impacts to a stronger focus on adaptation and adaptive capacity (Füssel & Klein, 2006).

Notwithstanding the lack of commonly agreed definitions and approaches, there is a great need to be able to assess and compare the vulnerability of regions, countries and sectors. Knowledge of vulnerability would enable scientists and decision-makers to anticipate and act on the adverse consequences of current and future changes, including those resulting from sea-level rise and other effects of climate change. Comparability is key to the notion of vulnerability: decision-makers are often interested in knowing which countries, regions, communities or sectors are most vulnerable, so that they can prioritise their activities.

How could a methodology for assessing vulnerability be specific enough to consider the unique circumstances of a given system while being generic enough to ensure that the vulnerability of this system can be compared with that of other systems, possibly assessed using different methodologies? In this chapter, we argue that this would require two elements: (i) a common domain-independent conceptual framework of vulnerability and (ii) a well-defined process that specifies how the frame-work's general concepts can be specialised to accommodate the specific case of the assessment.

A *common conceptual framework* is needed to enable unambiguous communication about vulnerability and meaningful comparison between vulnerability assessments. Given the diversity of types of natural and social systems under study, this common framework must be very general indeed: the definition of vulnerability should only include those elements that are absolutely necessary for avoiding ambiguity and it must be independent from a specific domain of application. A *well-defined process* is then needed to organise the specialisation of the framework's general concepts for the system of interest, resulting in a case-specific operational definition of vulnerability. This step requires detailed system understanding and the integration of expertise from different knowledge domains. Case-specific definitions of vulnerability cannot be prescribed, but the process of deriving them from the general concepts can be structured and facilitated.

This chapter presents a recent attempt at developing and applying these two elements. First, we present a general domain-independent formal framework of vulnerability proposed by Ionescu, Klein, Kavi Kumar, Hinkel, and Klein (2005) as an example of the first element. As an example of the second element we then present the Dynamic and Interactive Vulnerability Assessment (DIVA) method: a method developed by Hinkel (2005) to organise the iterative integration of knowledge and thus develop a case-specific operational definition of vulnerability. Finally, we show how the formal framework and the DIVA method have been applied in the EU-funded project DINAS-COAST (Dynamic and Interactive Assessment of National, Regional and Global Vulnerability of Coastal Zones to Climate Change and Sea-Level Rise) to develop the coastal vulnerability assessment tool DIVA (DINAS-COAST Consortium, 2006). Note that the chapter does not present results produced by the application of DIVA; these can be found in Nicholls, Klein, & Tol (Chapter 13).

The Evolution of Methodologies for Assessing Coastal Vulnerability

Before climate change emerged as an academic focus, vulnerability as such was not an important concept in coastal research. Traditionally, research in coastal zones has been conducted mainly by geologists, ecologists and engineers, roughly as follows (Klein, 2002):

- Geologists study coastal sedimentation patterns and the consequent dynamic processes of erosion and accretion over different spatial and temporal scales.
- Ecologists study the occurrence, diversity and functioning of coastal flora and fauna from the species to the ecosystem level.
- Engineers take a risk-based approach, assessing the probability of occurrence of storm surges and other extreme events that could jeopardise the integrity of the coast and the safety of coastal communities.

The challenge of climate change has spurred the collaboration between these three groups of coastal scientists; vulnerability has become an integrating focus of this research collaboration. Since 1990 a number of major efforts have been made to develop guidelines and methodologies for assessing coastal vulnerability to climate change, which combined the expertise of the three disciplines, complemented with economics.[1]

In 1992 the former Coastal Zone Management Subgroup of the IPCC published the latest version of its Common Methodology for Assessing the Vulnerability of Coastal Areas to Sea-Level Rise (IPCC CZMS, 1992). It comprises seven consecutive analytical steps that allow for the identification of populations and physical and natural resources at risk, and of the costs and feasibility of possible responses to adverse impacts. Results can be presented for the seven vulnerability indicators listed in Table 5.1.

The Common Methodology has been used as the basis of assessments in at least 46 countries; quantitative results were produced in 22 country case studies and eight subnational studies (for an overview see Nicholls, 1995). Hoozemans, Marchand, and Pennekamp (1993) applied the Common Methodology on a global scale. Studies that used the Common Methodology were meant to serve as preparatory assessments, identifying priority regions and priority sectors and providing an initial screening of the feasibility and effect of coastal protection measures. They have been successful in raising awareness of the potential magnitude of climate change and its possible consequences in coastal zones. They have thus provided a motivation for implementing policies and measures to control greenhouse gas emissions. In addition, they have encouraged long-term thinking and they have triggered more detailed local coastal studies in areas identified as particularly vulnerable, the results of which have contributed to coastal planning and management.

Nonetheless, a number of problems with the Common Methodology have been identified, which mainly concern its data intensity and its simplified approach to assessing

---

[1] Many involved in these efforts were unaware of the long history of vulnerability assessment in other disciplines, particularly the social sciences. In social-science research on poverty, food security and natural hazards, vulnerability is also interpreted in terms of potential harm and capacity to cope, but studies tend to focus in more depth on particular groups and communities within a society. In doing so, they take a quite different (i.e., bottom-up) approach to vulnerability assessment. This approach is typically place-based and cognisant of the rich variety of social, cultural, economic, institutional and other factors that define vulnerability. It does not rely on global or regional models to inform the analysis; instead the major source of information is the vulnerable community itself (Klein, 2002).

Table 5.1: The vulnerability indicators of the IPCC Common Methodology.

Indicator	Description
People affected	The people living in the hazard zone affected by sea-level rise
People at risk	The average annual number of people flooded by storm surge
Capital value at loss	The market value of infrastructure which could be lost due to sea-level rise
Land at loss	The area of land that would be lost due to sea-level rise
Wetland at loss	The area of wetland that would be lost due to sea-level rise
Adaptation costs	The costs of adapting to sea-level rise, with an overwhelming emphasis on protection
People at risk after adaptation	The average annual number of people flooded by storm surge, assuming the costed adaptation to be in place

biogeophysical and socio-economic system response (for a more detailed discussion see Klein & Nicholls, 1999). Alternative assessment methodologies have been proposed, but they have generally not been applied by anyone other than their developers. A semi-quantitative methodology proposed by Kay and Hay (1993) was applied in a number of South Pacific island countries, where it was felt that the Common Methodology put too much emphasis on market-based impacts. An index-based approach proposed by Gornitz, Daniels, White, and Birdwell (1994) included the risk of hurricanes and was developed for use along the east coast of the United States. However, it did not consider socio-economic factors.

The relative success of the Common Methodology led the IPCC to adopt its approach as a model for assessing the vulnerability of other, non-coastal systems to climate change. The top-down approach of the Common Methodology was intuitively attractive to the wider climate change community, whose work has been strongly model-orientated. In 1994, the IPCC published its Technical Guidelines for Assessing Climate Change Impacts and Adaptations (Carter, Parry, Nishioka, & Harasawa, 1994), which provide system-independent guidance to countries that wish to assess their vulnerability to climate change. The Technical Guidelines are outlined in a similar fashion to the Common Methodology, but fewer analytical steps are implied and less prior knowledge is assumed. In addition, the Technical Guidelines are not prescriptive in the choice of scenarios, tools and techniques to conduct the analysis. The United Nations Environment Programme (UNEP) Handbook on Methods for Climate Change Impact Assessments and Adaptation Strategies (Feenstra et al., 1998) offers a detailed elaboration of the IPCC Technical Guidelines for a range of socio-economic and physiographic systems, including coastal zones (Klein & Nicholls, 1998). The UNEP Handbook has been used in a number of developing countries under the UNEP Country Studies Programme and in the first phase of the Netherlands Climate Change Studies Assistance Programme. The United States Country Studies Program used similar guidance provided by Benioff, Guill, and Lee (1996).

In the late 1990s, the EU-funded project SURVAS (Synthesis and Upscaling of Sea-Level Rise Vulnerability Assessment Studies) aimed to synthesise and upscale all available coastal vulnerability studies and to develop standardised data sets for coastal impact indicators suitable for regional and global analysis (De la Vega-Leinert, Nicholls, & Tol, 2000; De La Vega-Leinert, Nicholls, Nasser Hassan, & El-Raey, 2000; see also http://www.survas.mdx.ac.uk/). However, this effort was only partially successful: synthesis and upscaling was impeded by the fact that studies had used different methodologies, scenarios and assumptions. As a result, until the publication of DIVA (DINAS-COAST Consortium, 2006) the global assessments by Hoozemans et al. (1993) and its updates by Baarse (1995) and Nicholls (2002, 2004) remained the only sources of global information on coastal vulnerability to sea-level rise.

A Formal Framework of Vulnerability

The first element required to assess vulnerability is a conceptualisation of vulnerability. We want this *a priori* conceptualisation to be as general as possible so that it can be applied to a variety of natural and social systems and it ensures comparability with other approaches. The formal framework proposed by Ionescu et al. (2005) serves these purposes. Their definition of vulnerability differs from most definitions in the literature in that it is independent from specific knowledge domains (i.e., scientific disciplines) and from the system of interest (e.g., a biological or social system). Vulnerability is defined on the basis of domain-independent mathematical concepts. In this chapter we only give a brief overview of the framework; for a full account see Ionescu et al. (2005).

The formal framework requires one to specify (i) the *entity* of which the vulnerability is assessed, (ii) the *stimulus* to which the entity would be vulnerable and (iii) the *preference criteria* that are used to evaluate the outcome of the interaction between the entity and the stimulus (e.g., an adverse or undesirable outcome). In other words, it is the vulnerability of an entity to a specific stimulus with respect to certain preference criteria. Examples in coastal zones include the vulnerability of Bangladesh to sea-level rise with respect to the number of people affected by coastal flooding, the vulnerability of tourist resorts in Florida to an increased intensity of hurricanes with respect to economic losses, the vulnerability of the Great Barrier Reef to increased sea-surface temperatures with respect to the degradation of coral ecosystems and the vulnerability of a fishing community in Vietnam to the conversion of mangroves into fishponds with respect to the loss of traditional livelihoods. Any definition of vulnerability must thus contain the three primitive concepts of entity, stimulus and preference criteria in order to convey meaningful information, and in fact most approaches described in the literature do (Brooks, 2003).

In the framework proposed by Ionescu et al. (2005), the entity of which the vulnerability is assessed is represented as a discrete dynamical system and the stimulus to which it is exposed is the system's exogenous input. The system's "reaction" to the exogenous input is given by

$$x^{k+1} = f(x^k, e^k) \qquad (5.1)$$

where $f: X \times E \rightarrow X$ is called the transition function of the system; X the set of states of the system; E the set of exogenous inputs to the system; and k the time step (a discrete system is considered).

The system's output is given by

$$y^k = g(x^k) \tag{5.2}$$

where $g: X \to Y$ is called the output function of the system and Y the set of outputs.

These outputs can be thought of as indicators of the state and are in general considered measurable or observable quantities.

The preference criteria are represented as a preorder $\leqslant$ on the set of outputs Y. A preorder is a reflexive and transitive relation and thus a very general mathematical model to represent preference criteria. The notation $y^k \leqslant y'^k$ means that the system that produces output y^k is considered to be "worse off" compared to the system that produces output y'^k.

The concepts introduced here now allow us to define vulnerability: a system is vulnerable to an exogenous input if it ends up "worse off" than it was before, or more formally:

Definition 1. A system (f, g) in state x^k is vulnerable to an exogenous input e^k with respect to preorder $\leqslant$ if and only if (iff) $y^{k+1} < y^k$.

In addition, it is possible to compare the vulnerability of one entity under different circumstances (i.e., in different states) or to another entity receiving the same exogenous input.

Definition 2. A system (f, g) is more vulnerable in state x^k than in state x'^k to an exogenous input e^k with respect to preorder $\leqslant$ iff

(i) the system in state x^k is vulnerable to exogenous input e^k with respect to preorder $\leqslant$,
(ii) $y^{k+1} \leqslant y'^{k+1}$.

Definition 3. A system (f, g) in state x^k is more vulnerable to an exogenous input e^k than a system (f', g') in state x'^k is to an exogenous input e'^k with respect to preorder $\leqslant$ iff

(i) it is vulnerable to e^k with respect to $\leqslant$,
(ii) $y^{k+1} \leqslant y'^{k+1}$.

While the concepts introduced so far have allowed us to define vulnerability, a further primitive concept is needed to include the notion of adaptation. Adaptation requires that the vulnerable entity has actions at its disposal to respond to the exogenous inputs it receives. To represent these actions, the dynamical system's transition function must be extended to include endogenous inputs:

$$x^{k+1} = f(x^k, e^k, u^k) \tag{5.3}$$

where u^k is an element of $U^k = U(x^k, e^k)$, the set of available endogenous inputs (or adaptation actions).

Adaptation involves choosing an *effective* action that will prevent the system from being worse off in the next time step (i.e., choose an action $u^k \in U^k$ such that *not* $(y^{k+1} < y^k)$) The size of the set of effective actions available to the system can be interpreted as the system's *adaptive capacity*.

Finally, we can define an *adaptation strategy* as a function that returns an adaptation action u^k for every state x^k of the system and for every exogenous input e^k it receives:

$$u^k = \phi(x^k, e^k) \tag{5.4}$$

where $\phi : X \times E \to U$.

A more elaborate description of the framework, along with examples, can be found in Ionescu et al. (2005).

The DIVA Method

The second element required to assess vulnerability is a well-defined process that organises how the formal framework's general concepts can be specialised to accommodate a specific case. This process involves two tasks. First, the mathematical concepts must be interpreted, that is, they must be mapped to components of the "real-world" system of interest (i.e., the vulnerable entity, the stimuli and the preference criteria). Second, the mathematical concepts must be specialised to represent their "real-world" counterparts. The mathematical forms of the state transition function (Eq. (5.1) or (5.3)), the output function (Eq. (5.2)) and the adaptation strategies (Eq. (5.4)) have to be specified in order to apply the framework's formal definitions. The product of this task is the operational definition.

The challenge of this process lies in the interdisciplinary nature of the two tasks, especially the second one. Knowledge from both the natural and the social sciences must be identified and integrated into a complete mathematical description of the system of interest. There is no single possible outcome when integrating knowledge into an operational definition of vulnerability. Different groups of experts tackling the same problem will inevitably come up with different specialisations and therefore with different definitions. Moreover, the definition of vulnerability evolves within a group of experts over the course of the assessment. The interactions between the various parts of the system are usually not fully understood at the start of an assessment; instead, such understanding is a result of the assessment itself.

How can an assessment methodology be designed to deal with the fact that the operational definition of vulnerability is almost certain to change as system knowledge develops over the course of the assessment? One way of dealing with vulnerability being a moving target is to design a methodology that allows for the development and refinement of operational definitions during the assessment. In other words, rather than to settle on *the* definition of vulnerability at the outset, an iterative process is agreed to develop and refine *good* definitions of vulnerability in response to the development of new, integrated knowledge.

Integrating knowledge can be particularly challenging when the participants in the assessment represent different scientific disciplines, use incompatible terminology and lack the time or funding for frequent face-to-face meetings. These challenges create a need for methods that foster the communication, collaboration and mutual learning between participants and thus lead to a better interdisciplinary understanding of the issues at hand and to a more adequate definition of vulnerability.

In DINAS-COAST, the DIVA method was developed specifically to facilitate the integration of knowledge from distributed experts with different disciplinary backgrounds. The

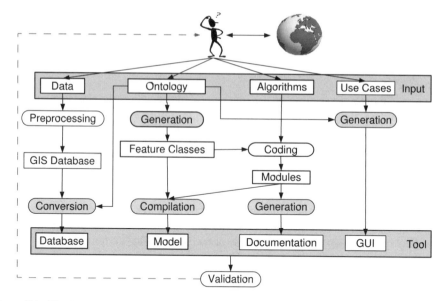

Figure 5.1: The DIVA development process. Boxes denote deliverables, ovals denote processes, shaded ovals denote automated processes.

DIVA method was then applied to build the DIVA tool (see next section on 'The Case of DINAS-COAST'). However, the DIVA method is a generic method for building modular integrated models by distributed partners and could be applied to any problem with similar requirements. For a detailed technical description of the DIVA method see Hinkel (2005).

The DIVA method consists of a modelling framework and a semi-automated development process. The modelling framework addresses the integration of knowledge at the product level: it frames the product (i.e., the DIVA tool) by providing a general *a priori* conceptualisation of the system to be modelled based on the formal framework presented in the earlier section on 'A Formal Framework of Vulnerability'. It does so by providing concepts for expressing static information about the system, as well as for representing the system's dynamics. The static information of the system is represented by a relational-data model consisting of geographical features, properties and relations. The geographical features represent the real-world phenomena such as rivers or countries. Properties capture the quantitative information about the features. For example, a country might have the property "area" or a river the property "length". Finally, relations describe how the features are structured. For example, a region might contain several countries. The dynamics of the system are represented in the form of differential equations, in accordance with the formal framework.

The development process then addresses the integration of knowledge at the process level. It organises the iterative specialising of the framework's general concepts to the needs of the specific problem addressed, thereby structuring the integration of knowledge from the various experts. Knowledge enters the process as four categories (see Figure 5.1):

- The system's ontology, which is a formal vocabulary for referring to properties of the modelled system.

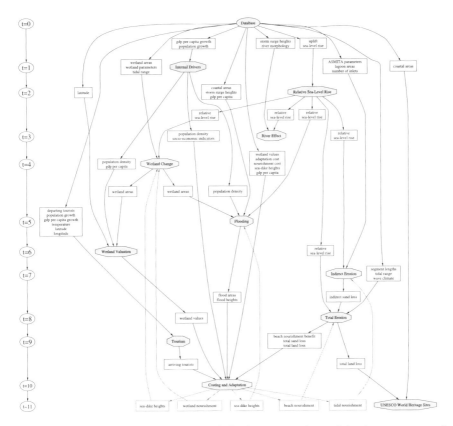

Figure 5.2: Module linkages in the DIVA model. Ovals represent the modules, boxes represent data, the solid arrows represent the flow of data during one time step and the dotted arrows represent the data fed into the next time step.

- The algorithms, which implement the system's state transition function, output function and adaptation strategies.
- The data, which express the initial state of the system and the inputs to which its vulnerability is being assessed (in the form of scenarios).
- The use-cases, which specify how the user can interact with the model via a graphical user interface (GUI).

The four categories of knowledge are interrelated; for example, new data may create the need to change existing algorithms or develop new ones with the consequent need to update the ontology. Once knowledge has entered the development process, most subsequent processes are automated. The development process can be iterated as many times as needed and while at any stage new knowledge can be incorporated, there is always a complete model available.

The first step of each iteration of the development process is the elaboration of a common formal vocabulary or ontology. The specific features, properties and relations that constitute the modelled system must be specified. All properties of the features must be classified into four categories according to the part they play in the system's dynamics: driver, state variable, diagnostic variable and parameter. For example, the country's area would most likely be static (a parameter), whereas its population might be driving the model. The compilation of the ontology is a joint responsibility of the project team.

In the next step the ontology is automatically translated into Java source code, which is then used by the project partners to code the algorithms. The hard coding in Java ensures that an algorithm only compiles if it is consistent with the ontology. Related algorithms are grouped into modules. For example, a social scientist could write a module called "CountryDynamics", which simulates how the properties of the feature "country" evolve over time. Before a module is submitted for inclusion into the integrated model it is run and validated in stand-alone mode.

The last step of each iteration of the development process consists of the analysis of the modules and their linkages, and the validation of the integrated model. The project website is automatically updated with every new submission of a module, offering documentation and the new model to download. An important document that is automatically generated is a graph visualising the data flow through the modules (Figure 5.2). With this graph the project team can analyse the interactions between the modules and decide whether any changes need to be made in the next iteration of the development process.

The main advantage of the iterative approach is that the specification of subsystem interfaces is not required before one can begin to develop and code the algorithms. This allows the module developers to take advantage of the interdisciplinary learning process that takes place over the course of the assessment.

The Case of DINAS-COAST

This section illustrates the methodological issues presented above with the help of an example. It shows how the formal framework and the DIVA method have been applied for the assessment of vulnerability of the world's coastal zone to sea-level rise within the project DINAS-COAST.

The Conceptualisation of Vulnerability

The vulnerable *entity* studied in DINAS-COAST is the world's coastline, or more specifically segments of it. To reflect its large natural and socio-economic diversity, the coastline was decomposed into segments that are assumed to be homogeneous in terms of vulnerability to sea-level rise but which vary in length, with an average length of 70 km. This segmentation was performed on the basis of physical, administrative and socio-economic criteria, producing 12,148 coastline segments in total (McFadden, Nicholls, Vafeidis, & Tol, 2006). Data on coastal characteristics needed for the calculation of potential impacts, such as coastal topography, population and protection status, are attributed to the coastline segments (Vafeidis, Nicholls, McFadden, Tol, Hinkel, Spencer, Grashoff, Boot, & Kelin, 2006; see also section on 'The Operational Definition: The DIVA Tool').

Characterising the vulnerable entity also includes identifying its potential adaptation actions and strategies. As will be discussed in more detail in the next section, four different adaptation strategies are considered. The project team's choice for these four strategies was motivated by a desire to provide users with the possibility to explore differences and trade-offs between strategies, as well as with the flexibility to define their own coastal protection standards. However, the coarse geographical scale of the analysis limits the usefulness of these results for coastal management.

The *stimuli* in DINAS-COAST that drive the assessment of vulnerability are scenarios of sea level, temperature, precipitation, coastal population and gross domestic product per capita. These scenarios are based on the four storylines of the IPCC Special Report on Emissions Scenarios (SRES; Nakićenović & Swart, 2000), which have been the standard source of scenarios for climate impact and vulnerability assessment for the past five years. More recently there has been increased interest in using scenarios for stabilising atmospheric greenhouse gas concentrations, but they were not available in time for use in DINAS-COAST.

The *preference criteria* relate to the output variables of the assessment. These largely correspond with the indicators used by the IPCC Common Methodology, listed in Table 5.1. This choice of output variables reflects both the desire to be able to compare between previous and current assessments and the fact that coastal research has focused strongly on these types of impact over the past decade. As a result of the latter factor, the assessment of impacts in the current version of the DIVA tool is considerably more sophisticated than the assessments carried out by Hoozemans et al. (1993), Baarse (1995) and Nicholls (2002, 2004). New output variables include effects on tourism arrivals and world heritage sites, as well as more detailed assessments of the costs of adaptation and the loss of wetlands, including their valuation.

The Operational Definition: The DIVA Tool

The conceptualisation of vulnerability above was developed into an operational definition of vulnerability by applying the DIVA method (section on 'The DIVA Method'). The product of this step is the DIVA tool, which consist of a global coastal database, a model, a set of scenarios and a GUI. The tool enables its users to simulate the effects of climate and socio-economic change and of adaptation on natural and human coastal systems on national, regional and global scales.

The vulnerable *entity* is represented by a computer model that implements and recursively applies the three functions of the formal framework: the state transition function f (Eq. (5.3)), the output function g (Eq. (5.2)) and the adaptation strategy ϕ (Eq. (5.4)). The complete model is a function h that, given an initial state x^0, takes a sequence of inputs $(e^1, e^2, \ldots, e^K)$ representing the evolution of the environment from time 1 to K and produces a sequence of outputs $(y^1, y^2, \ldots, y^k)$ representing the evolution of the coastal system:

$$(y^1, y^2, \ldots, y^K) = h((e^1, e^2, \ldots, e^K), x^0, \phi) \tag{5.5}$$

where $h: X \times E^K \times \Phi \rightarrow Y^K$, X is the set of states of the system; E the set of exogenous inputs; Φ the set of adaptation strategies; and Y the set of outputs of the system.

The transition function f and the output function g are unique, whereas the adaptation strategy ϕ can be selected from a set of possible strategies. As mentioned earlier, an adaptation strategy is a function that returns an adaptation action for each state of the system and for each input it receives (see Eq. (5.4)). The adaptation actions contained in the set of endogenous inputs U are (i) do nothing, (ii) build dikes, (iii) move away, (iv) nourish the beach, (v) nourish the tidal basins and (vi) nourish the wetlands. Some combinations of these actions are possible as well. For each time step (corresponding with model input) the DIVA model selects an adaptation action according to the following four adaptation strategies:

- *No adaptation.* the model only computes potential impacts.
- *Full protection.* raise dikes or nourish beaches as much as is necessary to preserve the *status quo* (i.e., x^0).
- *Optimal protection.* optimisation based on the comparison of the monetary costs and benefits of adaptation actions and potential impacts.
- *User-defined protection.* the user defines a return period against which to protect.

The functions f, g and ϕ are distributed across various modules. Each module represents a specific coastal subsystem and encapsulates the knowledge of one or more experts. Table 5.2 lists all the modules of the current version of the DIVA model (1.5.5) and Figure 5.2 shows the flow of data through the modules. The first modules to be invoked compute geodynamic effects of sea-level rise on coastal systems, including direct coastal erosion, erosion within tidal basins, changes in wetlands and the increase of the backwater effect in rivers. This is followed by an assessment of socio-economic impacts, either directly due to sea-level rise or indirectly via the geodynamic effects. The last module is the costing and adaptation module, which implements adaptation actions according to the user-selected strategy. These actions influence the calculations of the geodynamic effects and socio-economic impacts of the next time step.

The *stimuli* are represented by sea-level rise and socio-economic scenarios. Both sets of scenarios were developed to be mutually consistent on the basis of the four IPCC SRES storylines (Nakićenović & Swart, 2000). The climate scenarios were produced with the climate model of intermediate complexity CLIMBER-2 of the Potsdam Institute for Climate Impact Research (Petoukhov et al., 2000; Ganopolski et al., 2001), while the socio-economic scenarios were produced by the Hamburg University. The climate and socio-economic scenarios have been regionalised so as to allow for more realistic assessments. The climate scenarios have been made available for low-, medium- and high-climate sensitivities, which allows users to assess the range of possible impacts and their sensitivity to the climate scenarios. Although scenarios of temperature and precipitation have been developed, DIVA 1.5.5 makes limited use of them. This largely reflects the uncertainty surrounding the contribution of these climate variables to coastal vulnerability.

The *preference criteria* on the model's output have only partially been implemented. The DIVA model does not produce a scalar indicator of vulnerability. The model's output (Table 5.3) has many components and no preorder is given on the set of outputs Y. However, since the output is quantitative, a preorder (i.e., a total order) is given naturally on each component of the output. The monetary components are directly comparable and are used to calculate the "optimal protection" adaptation strategy. The non-monetary components have not been made comparable through normalisation. Rather, it is left to the user

Table 5.2: The modules of the DIVA model.

Module name	Author(s)	Description
Internal drivers	Richard Tol	Produces socio-economic scenarios
Relative sea-level rise	Robert Nicholls, Loraine McFadden, Jochen Hinkel	Creates relative sea-level rise scenarios by adding vertical land movement to the climate-induced sea-level scenarios
River effect	Rob Maaten	Calculates the distance from the river mouth over which variations in sea level are noticeable
Wetland change	Loraine McFadden, Jochen Hinkel	Calculates area change due to sea-level rise for six types of wetlands, taking into account the effect of flood defences
Flooding	Robert Nicholls, Jochen Hinkel	Calculates flooding due to sea-level rise and storm surges, taking into account the effect of flood defences
Wetland valuation	Luke Brander, Onno Kuik, Jan Vermaat	Calculates the value of different wetland types as a function of GDP, population density and wetland area
Indirect erosion	Luc Bijsterbosch, Zheng Bing Wang, Gerben Boot	Calculates the loss of land, the loss of sand and the demand for nourishment due to indirect erosion in tidal basins. This is a reduced version of the Delft Hydraulics ASMITA model (Stive, Capobianco, Wang, Ruol, & Buijsman, 1998)
Total erosion	Robert Nicholls	Calculates direct erosion on the open coast based on the Bruun rule. Adds up direct erosion and indirect erosion for the open coast, including the effects of nourishment where applied
Tourism	Richard Tol	Calculates number of tourists per country
Costing and adaptation	Richard Tol, Gerben Boot, Poul Grashoff, Jacqueline Hamilton, Jochen Hinkel, Loraine McFadden, Robert Nicholls	Calculates socio-economic impacts and either user-defined or optimal adaptation
World heritage sites	Richard Tol	Calculates whether a UNESCO world heritage site is threatened by sea-level rise

Table 5.3: Selected output of the DIVA model.

Issue	Indicator
Erosion	Land lost, sand lost in tidal basins
Flooding	Dike height, people at risk, people actually flooded
Saltwater intrusion	Area influenced by seawater intrusion into rivers
Wetlands	Area of six different types of wetlands, monetary value of wetlands
Costs	Adaptation cost, cost of nourishment, cost of building dikes, cost of saltwater intrusion, cost of migration, residual damage

to explore and compare the outputs that are produced by choosing different adaptation strategies and scenarios. This is facilitated by the GUI, which allows for the visual comparison of the outputs for different regions, time steps, scenarios and adaptation strategies in the form of graphs, tables and maps.

Conclusions and Outlook

This chapter has argued that the methodological advancement of vulnerability assessment would benefit from the development of two elements: (i) a domain-independent conceptual framework of vulnerability to enable unambiguous communication about vulnerability and meaningful comparison between vulnerability assessments and (ii) a process to organise the specialisation of the framework's general concepts into operational, system-specific definitions so as to facilitate the integration of knowledge from different experts and disciplines.

The formal framework proposed by Ionescu et al. (2005) is an example of the first element. The general conceptualisation can be applied to any system whose components can be mapped to the three primitive mathematical concepts of entity, stimulus and preference criteria. This is particularly useful when knowledge about natural and social systems needs to be integrated, as is the case when assessing coastal vulnerability. To take the same general starting points for different assessments ensures comparability. In addition, a formal framework is a prerequisite for computational approaches such as the one taken by DINAS-COAST.

The DIVA method is an example of the second element, a process that specifies how general concepts can be specialised into an operational definition to accommodate a specific case. It is an innovative method for developing an integrated model by geographically distributed partners, providing scientists with different backgrounds with a methodological procedure to harmonise their conceptualisations of the system of interest and with an intuitive interface to express and integrate their knowledge about it. The process of model development is well defined and automatically documented. As a result, the *status quo* is always available on the Internet, providing a basis for efficient communication and collaboration between project partners.

The generic nature of both the formal framework and the DIVA method makes them easily extensible and transferable to address new challenges, including non-coastal ones.

Improvements on the current version of the DIVA model could include developing a module for coral reefs and atolls, considering consequences of climate change other than sea-level rise (including extreme events), focusing more strongly on river coast interactions, refining the adaptation module and increasing the spatial resolution of the model, thus increasing DIVA's usefulness to coastal management. In addition, it is conceivable to develop regional versions of the DIVA tool, such as a DIVA-Europe or a DIVA-South Asia.

Acknowledgements

This chapter is based on research that has been funded by the Research Directorate-General of the European Commission through its projects DINAS-COAST (contract number EVK2-CT-2000-00084) and NeWater (contract number 511179-GOCE), the Deutsche Forschungsgemeinschaft (grant KL 611/14) and the START Visiting Scientist Program. The authors thank Cezar Ionescu, K.S. Kavi Kumar, Rupert Klein, Paul Flondor, Tom Downing and Poul Grashoff for stimulating discussions and valuable insights.

References

Adger, W. N. (2006). Vulnerability. *Global environmental change: Human and Policy Dimensions,* in press.

Baarse, G. (1995). *Development of an operational tool for global vulnerability assessment (GVA): Update of the number of people at risk due to sea-level rise and increased flooding probability.* CZM Centre Publication 3, Ministry of Transport, Public Works and Water Management, The Hague, The Netherlands, 17pp.

Benioff, R., Guill, S., & Lee, J. (Eds). (1996). *Vulnerability and adaptation assessments: An international handbook. Version 1.1* (xx+191pp. + Appendices). United States Country Studies Management Team, Dordrecht, The Netherlands: Kluwer Academic Publishers.

Brooks, N. (2003). *Vulnerability, risk and adaptation: A conceptual framework.* Working Paper 38, Tyndall Centre for Climate Change Research, University of East Anglia, Norwich, UK, 16pp.

Carter, T. R., Parry, M. L., Nishioka, S., & Harasawa, H. (Eds). (1994). *Technical guidelines for assessing climate change impacts and adaptations.* Report of Working Group II of the Intergovernmental Panel on Climate Change, University College London, UK and Centre for Global Environmental Research, Tsukuba, Japan, x+59pp.

De la Vega-Leinert, A. C., Nicholls, R. J., Nasser Hassan, A., & El-Raey, M. (Eds). (2000). *Proceedings of the SURVAS expert workshop on African vulnerability and adaptation to accelerated sea-level rise.* Cairo, Egypt, 5–8 November 2000, Flood Hazard Research Centre, Middlesex University, Enfield, UK, vi+104pp.

De la Vega-Leinert, A. C., Nicholls, R. J., & Tol, R. S. J. (Eds). (2000). *Proceedings of the SURVAS expert workshop on European vulnerability and adaptation to accelerated sea-level rise.* Hamburg, Germany, 19–21 June, Flood Hazard Research Centre, Middlesex University, Enfield, UK, viii+152pp.

DINAS-COAST Consortium. (2006). *DIVA 1.5.5.* Potsdam Institute for Climate Impact Research, Potsdam, Germany, CD-ROM.

Feenstra, J. F., Burton, I., Smith, J. B., & Tol, R. S. J. (Eds). (1998). *Handbook on methods for climate change impact assessment and adaptation strategies.* Version 2.0, United Nations Environment Programme and Institute for Environmental Studies, Vrije Universiteit, Nairobi, Kenya and Amsterdam, The Netherlands, xxvi+422 pp.

Füssel, H.-M., & Klein, R. J. T. (2006). Climate change vulnerability assessments: An evolution of conceptual thinking. *Climatic Change, 75*(3), 301–324.

Ganopolski, A., Petoukhov, V., Rahmstorf, S., Brovkin, V., Claussen, M., Eliseev, A., & Kubatzki, C. (2001). CLIMBER-2: A climate system model of intermediate complexity. Part II: Model sensitivity. *Climate Dynamics, 17*, 735–751.

Gornitz, V. M., Daniels, R. C., White, T. W., & Birdwell, K. R. (1994). The development of a coastal risk assessment database: Vulnerability to sea-level rise in the U.S. Southeast. *Journal of Coastal Research*, special issue, *12*, 327–338.

Hinkel, J. (2005). DIVA: An iterative method for building modular integrated models. *Advances in Geosciences, 4*, 45–50.

Hoozemans, F. M. J., Marchand, M., & Pennekamp, H. A. (1993). *Sea level rise: A global vulnerability assessment. Vulnerability assessments for population, coastal wetlands and rice production on a global scale* (2nd revised ed., xxxii+184pp.). Delft and The Hague, The Netherlands: Delft Hydraulics and Rijkswaterstaat.

Ionescu, C., Klein, R. J. T., Kavi Kumar, K. S., Hinkel, J., & Klein, R. (2005). *Towards a formal framework of vulnerability to climate change*. NeWater Working Paper 2 and FAVAIA Working Paper 1, Potsdam Institute for Climate Impact Research, Potsdam, Germany, ii+20pp.

IPCC CZMS. (1992). A common methodology for assessing vulnerability to sea-level rise. Second revision. In: *Global climate change and the rising challenge of the sea*, Report of the Coastal Zone Management Subgroup, Response Strategies Working Group of the Intergovernmental Panel on Climate Change, Ministry of Transport, Public Works and Water Management, The Hague, The Netherlands, Appendix C, 27pp.

Kay, R. C., & Hay, J. E. (1993). A decision support approach to coastal vulnerability and resilience assessment: A tool for integrated coastal zone management. In: R. F. McLean & N. Mimura (Eds), *Vulnerability assessment to sea-level rise and coastal zone management*. Proceedings of the IPCC/WCC'93 Eastern Hemisphere Workshop, Tsukuba, Japan, 3–6 August, Department of Environment, Sport and Territories, Canberra, Australia, pp. 213–225.

Klein, R. J. T. (2002). *Coastal vulnerability, resilience and adaptation to climate change: An interdisciplinary perspective*. Ph.D. thesis, Christian-Albrechts-Universität zu Kiel, Germany, ix+133pp.

Klein, R. J. T., & Nicholls, R. J. (1998). Coastal zones. In: J. F. Feenstra, I. Burton, J. B. Smith & R. S. J. Tol (Eds), *Handbook on methods for climate change impact assessment and adaptation strategies*, (pp. 7.1–7.35). Version 2.0, United Nations Environment Programme and Institute for Environmental Studies, Vrije Universiteit, Nairobi, Kenya and Amsterdam, The Netherlands.

Klein, R. J. T., & Nicholls, R. J. (1999). Assessment of coastal vulnerability to climate change. *Ambio, 28*(2), 182–187.

McCarthy, J. J., Canziani, O. F, Leary, N. A., Dokken, D. J., & White K. S. (Eds). (2001). *Climate change 2001: Impacts, adaptation and vulnerability*. Contribution of Working Group II to the Third Assessment Report of the Intergovernmental Panel on Climate Change, Cambridge University Press, Cambridge, UK, x+1032pp.

McFadden, L., Nicholls, R. J., Vafeidis, A. T., & Tol, R. S. J. (2006). A methodology for modelling coastal space for global assessment. *Journal of Coastal Research*, in press.

Nakićenović, N., & Swart, R. (Eds). (2000). *Emissions scenarios*. Special Report of Working Group III of the Intergovernmental Panel on Climate Change, Cambridge University Press, Cambridge, UK, x+599pp.

Nicholls, R. J. (1995). Synthesis of vulnerability analysis studies. In: P. Beukenkamp et al. (Eds), *Proceedings of the World Coast Conference 1993*, Noordwijk, The Netherlands, 1–5 November 1993, Coastal Zone Management Centre Publication 4, The Hague, The Netherlands: National Institute for Coastal and Marine Management (pp. 181–216).

Nicholls, R. J. (2002). Analysis of global impacts of sea-level rise: A case study of flooding. *Physics and Chemistry of the Earth, Parts A/B/C, 27*(32–34), 1455–1466.

Nicholls, R. J. (2004). Coastal flooding and wetland loss in the 21st century: Changes under the SRES climate and socio-economic scenarios. *Global Environmental Change: Human and Policy Dimensions, 14*(1), 69–86.

O'Brien, K., Eriksen, S. Schjolden, A., & Nygaard, L. (2004). *What's in a word? Conflicting interpretations of vulnerability in climate change research*. Working Paper 2004:04, Center for International Climate and Environmental Research, University of Oslo, Oslo, Norway, iii+16pp.

Petoukhov, V., Ganopolski, A., Brovkin, V., Claussen, M., Eliseev, A., Kubatzki, C., & Rahmstorf, S. (2000). CLIMBER-2: A climate system model of intermediate complexity. Part I: Model description and performance for present climate. *Climate Dynamics, 16*, 1–17.

Stive, M. J. F., Capobianco, M., Wang, Z. B., Ruol, P., & Buijsman, M. C. (1998). Morphodynamics of a tidal lagoon and the adjacent coast. In: J. Dronkers & M. Scheffers (Eds), *Physics of estuaries and coastal seas* (pp. 397–407). Rotterdam, The Netherlands: Balkema.

Vafeidis, A. T., Nicholls, R. J., McFadden, L., Tol, R. S. J., Hinkel, J., Spencer, T., Grashoff, P. S., Boot, G., & Klein, R. J. T. (2006). A new global coastal database for impact and vulnerability analysis to sea-level rise. *Journal of Coastal Research*, under review.

Chapter 6

The Vulnerability and Sustainability of Deltaic Coasts: The Case of the Ebro Delta, Spain

Augustin Sánchez-Arcilla, Jose A. Jiménez and Herminia I. Valdemoro

Introduction

Deltaic systems are geomorphic features resulting from interacting river and marine dynamics and are subject to intense geophysical and anthropogenic processes. In essence riverine 'drivers' supply water, sediment and nutrients while marine 'drivers' reshape and disperse these inputs. Gravity and consolidation lead to subsidence while human activities lead to demands for hard and fixed engineered defences of the deltaic body and values, which are inherently contrary to the natural dynamic properties of these systems.

Because of this set of circumstances, deltaic systems are particularly vulnerable to any change in the factors that drive their evolution and, generally, they are already experiencing relative sea level rise (RSLR) due to their own subsidence, regardless of any general or regional eustatic trend. They also generally suffer from accelerated sediment starvation due to widespread river regulation and damming in their catchments (e.g. Cencini, 1998; Chen & Zong, 1998; Stanley & Warne, 1998; Woodroffe et al., in press).

This situation is well illustrated by the Ebro delta (Figure 6.1) located on the Spanish Mediterranean coast 200 km south of Barcelona. It is a relatively 'simple' delta, compared to other large deltas in the Mediterranean such as the Rhone or Po, with an outer coastline of about 50 km and a sub-aerial extent of about 320 km^2. The extensive damming in the Ebro catchment has radically reduced the overall sediment supply to the deltaic system which has therefore evolved from experiencing accretion to show stability in terms of its sub-aerial surface, although still experiencing strong reshaping processes (e.g. Jiménez & Sánchez-Arcilla, 1993; Guillén & Palanques, 1997; Jiménez et al., 1997).

The Ebro delta is therefore a particularly vulnerable coastal environment. The vulnerability concept has been used recently with different meanings, leading to different results (e.g. Klein & Nicholls, 1999; Kelly & Adger, 2000). In this chapter we define it as the

Managing Coastal Vulnerability
Copyright © 2007 by Elsevier Ltd.
All rights of reproduction in any form reserved.
ISBN: 0-08-044703-1

Figure 6.1: The Ebro delta.

characteristic of a system that describes its potential to be harmed (Gouldby & Samuels, 2005). It can be applied to the whole deltaic system or to any of the different components within it, that is the physical, ecological or socio-economic sub-systems. Within this context, deltaic vulnerability can be defined as the potential of the deltaic system or sub-system considered to be harmed (e.g. Sánchez-Arcilla et al., 1998). This 'potential' is normally evaluated as the expected value of the damage the system may experience and, in consequence, it embraces both the biophysical and the human dimensions. In this sense, the concept of deltaic vulnerability assumed here departs from some existing approaches in which vulnerability is defined only in terms of the human dimension (e.g. Blaikie et al., 1994).

According to our definition, vulnerability can be evaluated using any unit of measurement able properly to represent/quantify the sub-system or system of interest. Thus, the geomorphic vulnerability of the system can be assessed in terms of sub-aerial surface changes (Sánchez-Arcilla et al., 1998) and/or vertical elevation rates (Day et al., 1997). This geomorphic component refers to the physical sub-system hosting the human and natural values of the delta.

If we move from the concept of vulnerability to that of sustainability, we need a deltaic sustainability concept which makes reference to the delta's functioning. Therefore, sustainability should refer to the balance between the damage potential and the recovery potential that a coastal deltaic unit will experience at a given timescale, be it natural or human induced. It should also be definable for a component of the deltaic system and with reference to a specific geographical or spatial scale. Day et al. (1997) thus applied this concept to the geomorphic (only for the deltaic plain), ecological and economic deltaic sub-systems.

Based on these antecedents, the aim of this chapter is to present a framework with which to analyse deltaic vulnerability — and sustainability — due to coastal changes at different scales. The goal is to assess the usefulness of vulnerability assessment as a tool for deltaic coastal zone management. To do this, we present two parallel approaches. The first is flux-based and vulnerability is here quantified in terms of a balance equation which is applied to the overall deltaic system. The second approach is based on the evaluation of the different deltaic functions. This implies and involves assessing the influence of coastal processes on the relevant deltaic system functions and it requires an appropriate unit of measurement/quantification for each function. The analysis of both approaches is formulated from a coastal morphodynamics standpoint and it is illustrated with information from the Ebro delta.

A Flux-Based Assessment of Vulnerability for the Geomorphic Component

This approach makes reference to the use of a unit of measurement for assessing deltaic vulnerability in which the time dimension is explicitly included. To properly apply it, the different contributors (positive and negative) to the system's vulnerability have to be identified and quantified. Thus, instead of analysing the overall integrated system behaviour at a given timescale, the assessment is obtained by accounting for the different fluxes at that scale. In schematic terms, this approach can be simply formulated as:

$$\Psi_\chi = \sum \Psi_i^+ + \sum \Psi_j^- \tag{6.1}$$

where Ψ_χ is the variable used to quantify the vulnerability of the selected sub-system or component χ, Ψ^+ are the positive contributions to such a value and Ψ^- the negative ones, all of them in terms of their respective units per time.

When this is applied to the geomorphic component of a delta, such an equation is representing the continuity equation applied to the volume of sediment in the delta. The different contributions become the sediment fluxes associated to all the driving agents acting on the deltaic body at the selected timescale (Sánchez-Arcilla & Jiménez, 1997).

Once this approach has been selected, deriving a value for the system's vulnerability requires establishing a reference state and a given timescale. This is especially necessary in the case of deltas since they are by definition accretionary geomorphic features and, if

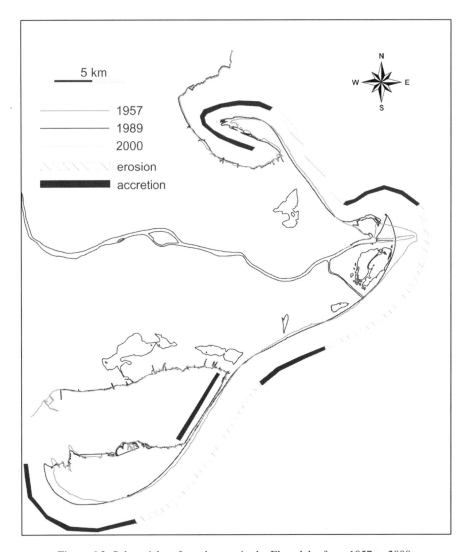

Figure 6.2: Sub-aerial surface changes in the Ebro delta from 1957 to 2000.

we compare any present delta in the world with their former stages and states, all of them will show 'worsening' conditions.

The integrated effects of this balance equation for the Ebro delta are shown in Figure 6.2 in terms of the net changes in its sub-aerial deltaic surface from 1957 until 2000. The sequence starts with the 1957–1973 period which is the first one with reliable aerial information for characterizing the deltaic behaviour prior to the construction of large dams in the drainage basin. In this case, the integrated effect of sediment fluxes has been quantified in m^2, which means working in surface units instead of the volumetric units that should be used to assess sediment fluxes. However there are merits in the surface measurement approach as this kind of analyses is mainly oriented to land managers and similar stakeholders, and their main concern is usually expressed in terms of the affected surface of the land and not in terms of volume of sediment to be lost.

The results obtained from this type of analysis show that the Ebro delta's sub-aerial surface area was experiencing some slight change during this period, with a net increase from 1957 of about 40 ha. This area is, however, below the accuracy of the employed data, and thus we should infer a quasi-equilibrium stage in terms of the delta's sub-aerial surface. Although this could be used to argue that the Ebro delta system is not 'in danger' (or vulnerable) at this scale, when this behaviour is compared to the pre-dam situation in which significant deltaic surface increase was the predominant characteristic (Guillén, 1992; Jiménez, 1996), we can see a clear change in the evolutionary trend at a century-based timescale.

Moreover, when these results are presented in an integrated manner (i.e. as the net surface change), they can hide part of the system's real behaviour and even 'mask' the true geomorphic vulnerability of the system. Figure 6.3 shows the negative and positive contributions to the sub-aerial surface changes in the Ebro delta for the same period. As we can see, although the net changes are very low, the 'gross' changes are still significant since the delta's coastline is significantly reshaped during this period (Jiménez et al., 1997): some coastal stretches are retreating while others are accreting (Figure 6.2).

Further insight that can be gleaned from Figure 6.3 is that the deltaic surface changes seem to be attenuating with time, which is consistent with coastal systems approaching a theoretical equilibrium state when subjected to steady driving forces. However these results reflect the timescale taken, and are based only on the sub-aerial deltaic surface area. In order to properly interpret these results, we need to note that most of accretion zones in the Ebro deltaic coast are shallow water areas whereas the erosion stretches are on the open coast. As a consequence, to achieve a gain of a given surface area, for example, in the northern spit (Fangar) or the river mouth area (Garxal), the required sediment volume

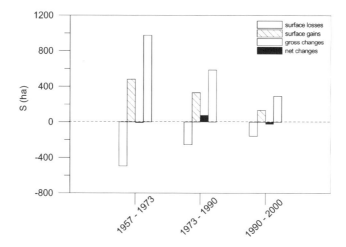

Figure 6.3: Area changes around the shoreline of the Ebro delta from 1957 to 2000.

is smaller than the amount 'produced' in the erosion of the same surface area in the easternmost apex (Cap Tortosa). This implies that although the Ebro delta can be considered at equilibrium in terms of its sub-aerial surface, this is not necessarily the case with regard to its sediment volume.

With this approach, the flux-based sustainability for the geomorphic component can be easily obtained from the net flux of sediment in a given time interval and for a given coastal unit. Thus, the delta is considered to be sustainable if overall input fluxes of sediment are equal or larger than outward fluxes. This can be further specified for the coastal fringe and/or the deltaic plain, taking into account the different implications of this geomorphic sustainability. The deltaic plain can be considered sustainable if vertical accretion equals or exceeds the rate of RSLR (Baumann et al., 1984; Day et al., 1997) whereas the deltaic coastal fringe can be considered as sustainable, in geomorphic terms, if the sediment balance is equal to or greater than zero. In terms of sub-aerial surface changes this means a state so as to maintain or increase its net value (Sánchez-Arcilla et al., 1998) This geomorphic sustainability can be written as:

$$\Psi_\chi = \sum \Psi_i^+ + \sum \Psi_j^- \geq 0 \tag{6.2}$$

When applied to the deltaic coast, this geomorphic sustainability equation results in an evaluation of the sediment budget in both horizontal dimensions. This must be done for all sedimentary fluxes resulting from the prevailing climatic conditions and the corresponding management policy within the time interval being considered:

$$S_{in} - S_{out} \geq 0 \tag{6.3}$$

where S_{in} are input sediment fluxes (gains) and S_{out} are output sediment fluxes (losses). In this regard Sánchez-Arcilla and Jiménez (1997) detail the main coastal processes (and associated sediment fluxes) controlling the coastal geomorphic sustainability of the Ebro delta coast at the long-term scale.

According to Jiménez (1996), and consistent with results shown in Figure 6.3, the overall sediment budget for the Ebro delta coast is in quasi-equilibrium, that is to say the gains and losses appear to be in balance, within the available levels of measurement accuracy. It has to be stressed that this balance refers to sandy sediments, as those contributing to the maintenance of the deltaic coast, whereas fine sediments are exported off-shore to the shelf. Applying the above established criteria results in a deltaic coastal fringe which is geomorphically sustainable in terms of its sub-aerial surface at a decadal timescale. However this is also reflects the situation whereby the actual delta is in a delicate state of balance between formation and reduction processes.

This has serious implications for the geomorphic sustainability in the sense that deltaic formation processes are mainly controlled or affected by human actions in the drainage basin and, under the present regime, it is not realistic to assume any improvement or change from present conditions. This is because the river regulation policy affecting potential sediment supplies is essentially controlled by hydroelectric authorities' regulating reservoirs in the river basin. On the other hand, the deltaic reduction processes — mainly controlled by wave action and sea level rise — are likely to be affected by climate change and available projections seem to forecast an increase in the intensity of those processes (Houghton et al., 2001).

The delta is therefore being 'squeezed' between a static landward situation that denies it its sediment lifeblood, and a dynamic shoreline situation that generates strong erosive forces.

Finally, however, we need to appreciate that in this approach, negative contributions are counteracted and counterbalanced by positive ones. In fact, the use of net changes or a flux-based balance implicitly assumes that surface gains and losses have equal consequences for the delta (see the different contributions to the Ebro delta surface changes in Figure 6.2). However, from a manager/stakeholder standpoint this need not necessarily be true and, in fact, in most cases it is not. In consequence, it is highly desirable to consider separately areas subjected to erosion and accretion. This disaggregation of the deltaic behaviour is needed when we assume that a part of the area is playing a given function that does not necessarily have a homogeneous spatial distribution. The practical consequence of this is to move from an analysis based on an overall delta vulnerability assessment to one in which the vulnerability of the different deltaic functions is quantified, as below.

A Functionally Based Assessment of Vulnerability

When an assessment is based on the system's functions, we have to quantify separately the vulnerability of the deltaic system for each of those main functions.

If we consider the delta as an ecosystem, the potential functions to be affected by any change in the state of the system could be selected from those listed by de Groot (2006). To make the analysis more user-oriented, especially from a manager/stake-holder standpoint, it would be more convenient to use the equivalent 'goods and services' that the functions produce. In this work, and as a first approximation, two main categories of functions/goods and services have been selected:

(i) *Natural* — which characterises the role of the delta in supporting natural areas of high-ecological value with their habitat functions, recreation functions related to ecotourism, and functions for science and education, etc.
(ii) *Cultivation* — which characterises the role and function of the delta in supplying land for agriculture which is the main economic activity in the area.

One possible approach to analysing the deltaic vulnerability for each of these functions would be to evaluate how a function is affected by a given system response. However, this would imply a full understanding of the system's functioning, including all feedbacks, and being able accurately to model such relationships. Instead of this, in this work we have approached the problem in a parametric way, such that instead of evaluating the direct effect of the involved use or resource, we just evaluate the effect — positive or negative — of the delta's surface supplying such a service (i.e. supporting the function). This territory-based approach requires a lower level of knowledge since it 'just' requires knowing the spatial distribution of each analysed function and the spatial extent of the potential resource effect.

Natural Function

As mentioned before, although the overall deltaic surface is not experiencing a measurable decrease at decadal timescale, coastline evolution determines the presence of erosive

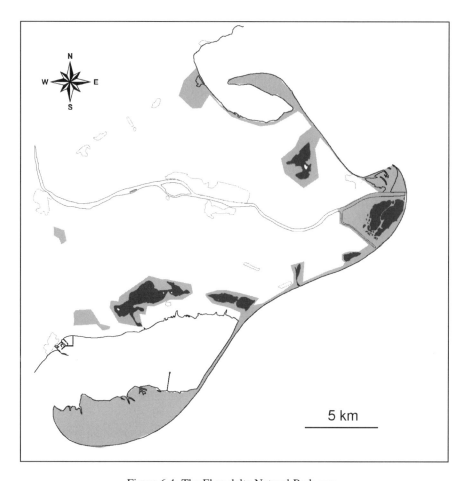

Figure 6.4: The Ebro delta Natural Park area.

and accretive areas with locally significant surface changes (Figure 6.3). To evaluate the deltaic vulnerability regarding the *natural* function due to these coastline changes we have to carry out a spatial analysis of the specific natural values of each surface area. The vulnerability assessment of the natural function of the delta leads us to consider that long-term coastline changes are the main threat, because the most valued natural areas of the Ebro delta, including all those protected by legislation (e.g. the Natural Park), are adjacent to the coastline (Figure 6.4). As a consequence, shoreline changes will have a definite influence (positive or negative) on such functions.

As an example, the area experiencing the largest erosional impact is the Illa de Buda whose hinterland is occupied by Els Calaixos, a wetland of about 800 ha, of which about 350 ha is made up of a central lagoon and the remaining part comprises salt marshes, reed swamps, rushes and patches of *Salicornia* communities. Hence, coastal erosion in this area will affect the deltaic natural functions by modifying the values associated to this type of ecosystem (Valdemoro et al., in press). To account for this loss of natural value we have to specify the affected surface area and the type of ecosystem to be lost. To put this in context, according to Menéndez et al. (2002) this is the only remaining lagoon in the Ebro delta with characteristics that allow the growth of freshwater and saltwater associated primary producers. This means that, although, there are other wetlands in the Ebro delta, there is no other wetland with the same properties/ structure.

Deltaic accretive areas where most of the sediment is deposited and the sub-aerial surface grows significantly are represented by the two spits. In both spits, the Banya at the south and El Fangar at the north, the expansion areas are basically beaches where the deposit of sediment creates wide low-lying sand surfaces, easily inundated by small water level variations. They form ideal habitats for a large part of the bird population of the delta. In this case, to account for this gain in natural value, as a result of the accretion of sediment, we have to specify the increase in surface area and the type of biotope gain.

Of course, regardless of the intrinsic value of each biotope, they have very different characteristics and they cannot be counted sensibly in a vulnerability analysis unless some form of standardisation of their values can be applied. Examples could be the net primary productivity (e.g. Cardoch et al., 2002) or any other method converting these values to economic ones (e.g. Costanza et al., 1997). This prevents the direct use of the affected overall sub-aerial surface for estimating vulnerability and, because of this, we have solved the problem by separately accounting for deltaic surface changes as a function of their individual characteristics.

Table 6.1 shows the main deltaic surface changes affecting natural functions due to coastline reshaping from 1957 to 2000. To take into account the different quality of the natural functions we have selected different units according to their natural characteristics: *Buda Island* refers to the wetland described above where fresh water and saltwater primary producers coexist, *Garxal* is the shallow water area developed in the river mouth (one of the few expansion areas of the delta), *sandy spits* refers to the sandy natural environments in both spits, *coastal wetlands* refers to the remaining wetland surface of the Ebro delta and *Natural Park* is the deltaic area protected under this under the Park's administration.

Table 6.1: Estimation of the natural (function based) partial vulnerability index for the Ebro delta due to long-term coastline changes from 1957 to 2000.

Biotope	**ΔS (ha)**	**S(ha)**
Buda Island	-150^a	800
Garxal	+168.5	185.7
Coastal wetlands	+18.5	2695^b
Sandy spits	+298.5	3098
Natural Park	+41.2	7802^b

Note: ΔS, surface change and S, surface area corresponding to the year 2000.
[a]Values indicates loss.
[b]Values have been obtained from images provided by the Environment Department of the Government of Catalonia.

Table 6.2: Potential impact (ratio of affected surface to actual surface) of different units contributing to the deltaic natural function from 1957 to 2000 for different weighting scales.

	Weighting scale				
	Buda Island	Garxal	Coastal wetlands	Sandy spits	Natural Park
Buda Island	−0.19[a]	n.a.	−0.056[a]	n.a.	−0.019[a]
Garxal	n.a.	0.91	0.063	n.a.	0.022
Costal wetlands	n.a.	n.a.	0.007	n.a.	0.024
Sandy spits	n.a.	n.a.	n.a.	0.19	0.038
Natural Park	n.a.	n.a.	n.a.	n.a.	0.005

Note: n.a., not applicable
[a]Values (negative) indicate loss and positive ones indicate surface gain.

Our index of *relative impact* — RI_χ — is calculated as the ratio of the affected surface supporting/supplying a given function, χ, with respect to the available surface (the scale of weighting). It can be written as:

$$RI_\chi = \frac{\Delta S_\chi}{S_\chi} \qquad (6.4)$$

where ΔS_χ is the measured or estimated deltaic surface change (depending on the type of analysis — hindcasting or forecasting) at a given temporal scale supporting the χ function, and S_χ is the total deltaic surface associated to such a function. Negative values indicate a vulnerable system (deltaic surfaces supporting/supplying a given function being lost) whereas positive values will indicate a resilient one (deltaic surfaces supporting/supplying a given function being gained).

Table 6.2 shows the values of the relative impact index calculated for each unit contributing to the natural function using different weighting values. The results obtained clearly show the importance of the weighting scale in this surface-based approach. Thus, if we consider the modifications suffered by the Buda Island ecosystems due to decadal-scale shoreline changes, a value of −0.19 is obtained (related to a maximum one of −1.0 which should mean total disappearance) when the impact is scaled to only those deltaic biotopes having the same properties (and, in consequence, with same natural values). However, if the same impact is scaled to the Natural Park extension (i.e. all deltaic areas comprising all the possible natural values protected by law), the same ecosystem change will give a value of −0.019, i.e. 10 times lower. This last assessment assumes that all the deltaic surface included under the Natural Park will have the same 'value' (i.e. function/service). This can clearly send a 'wrong' message to the managers, because a loss of wetlands should not be compensated for by a gain of sandy beaches since they will support very different ecosystems and, in consequence, they will supply different goods and services associated with their natural functions.

This emphasises the fact that when dealing with natural functions, unless they are all given or measured by using a standard measurement unit, or by considering changes in a single-ecosystem type (with an homogeneous spatial distribution of properties and structure along the coast), each ecosystem must be separately accounted for as surface gains and losses will not necessarily compensate each other. As a consequence, if the values of relative impact or effect introduced above are used to quantify the system's natural vulnerability, it is recommended that the selected weighting unit be the total surface area with similar natural characteristics or values.

Even following this recommendation, it is not clear how to interpret the values obtained on a vulnerability scale. The question for the manager would be: how vulnerable is the Buda Island regarding its natural function(s)? According to the proposed approach for weighting deltaic changes, the corresponding value is -0.19, which in essence means that 19% of the wetland surface (or its natural values) has been lost in the timescale of the analysis (from 1957 to 2000). To assess the importance of this loss we need to introduce some additional criteria taking into account the final goal of the assessment.

In this analysis we have used the number of units of a given ecological structure to be affected. Thus, although only 19% of the existing deltaic surface of this category is affected by long-term coastline changes there is only one physiographic unit with those values and it is threatened. This implies that 100% of such existing environments are in danger. If forcing conditions — natural or human induced — driving such results are steady, this would mean that the potential vulnerability of the system should be regarded as 100%. In other words, in the long-term perspective and considering the contribution of the Buda Island to the natural functions of the Ebro delta, that system should be classified as highly vulnerable.

This reasoning is also applicable to the concept of sustainability. In this case, the previous equation defining the requirements for a sustainable system should be adapted to

$$S_{gain} - S_{loss} \geq 0 \qquad (6.5)$$

where S_{gain} accounts for the increment in deltaic surface supporting the ecosystem of interest and S_{loss} accounts for surface losses. Applying this to the Buda Island case, we obtain a negative value $S_{gain} - S_{loss} = -150$ ha indicating that at this scale, and assuming that conditions controlling the coastal response will not vary, the system is not sustainable.

With this equivalence between deltaic surface and ecosystem values, we are assuming that net primary production values are homogeneously distributed in the physiographic unit of interest. Although this is not strictly true (e.g. see Menéndez et al. (2002) for a description in the cross-shore distribution of primary producers in the Buda Island), here we assume that coastal retreat and wetland reduction in the seaside border will be accompanied by a reconstruction of the ecosystem structure in this border as the shoreline migrates landwards. This means that the remaining wetland will have the same ecological properties as the original one. Of course, if we include any additional processes, such as those controlling water quality which could affect the ecosystem's health inside the lagoon, a more accurate analysis at a 'higher' level of knowledge could be performed.

Cultivation Function

To quantify the deltaic vulnerability from the point of view of the delta's *cultivation* function we have selected two different coastal morphodynamic processes that significantly influence the deltaic surface used for agriculture. As in the previous case, evaluating the effect of this deltaic function is done in an indirect manner, that is by measuring the deltaic surface dedicated to agriculture directly affected by the evolutionary processes that we are considering.

About 19,500 ha of the Ebro delta's surface is used for rice production, which is the main crop, covering about 85% of the cultivated area. This pattern arises for three main reasons (e.g. Casanova, 1998): (i) the presence of saline groundwater over most of the delta near the soil surface that it is counteracted by a continuous flow of irrigation water, (ii) market factors whereby the European Commission provides an intervention price (much higher than for other cereals but more stable than for vegetables) and (iii) environmental policies favoured by the European Commission.

Although the cultivated lands extend along the entire deltaic plain and some areas are very close to the beaches, it is not expected that this function will be directly affected by coastline changes at the decadal timescale. However, this does not mean that coastal processes will not affect it. Two coastal processes have thus been identified as relevant in affecting cultivation in the Ebro delta: (i) the impact of coastal storms (associated with the episodic event scale) and (ii) relative sea level rise (associated with the long-term scale).

Storm impacts on the Ebro delta coast usually occur with the coexistence of extreme high water levels due to the passage of low pressure systems off the Ebro delta coast and eastern wave storms (e.g. Jiménez et al., 1997). Under these conditions, waves are able to act on the usually non-exposed part of the coastal fringe and to produce large amounts of erosion in a very short-time period. Although the entire deltaic coast is subjected to the action of such events, the more vulnerable stretches are those with a narrow emerged beach and fronted by a low-crested bar or bar system. One of these sensitive coastal stretches, where the hinterland is occupied by rice fields, is the Marquesa beach at the Northern hemidelta (Figure 6.1).

Figure 6.5 shows an example of the impact of a storm in November 2001 on the Marquesa beach. This storm significantly eroded the entire coastline of the delta, but this area was one of the most affected. One of the processes occurring during the storm was the overtopping of the beach and the corresponding massive overwash of the rice fields close to the shoreline. This overwash directly affected the hinterland, where large volumes of sand were deposited, and, simultaneously, some roads and infrastructure located close to the shoreline were also affected.

A preliminary estimate from the Department of Agriculture after the storm impact in about 400 ha of rice fields of the affected area in the Northern hemidelta gave an estimated economic impact of €600,000 (Generalitat de Catalunya, 2004). This storm was the most powerful event recorded in the Ebro delta since the current wave buoy was deployed in 1990 (Jiménez et al., 2005), and in consequence it can be considered as a good estimate of the potential threat to the cultivated lands by storms in the area.

If we estimate the deltaic vulnerability regarding cultivation, based on the total cultivated rice area in the delta, and using this 400 ha as a measure of the area to be affected, a potential impact value of -0.02 would be obtained. In relative terms, this is rather small when compared with the total deltaic area dedicated to agriculture. However it will represent a serious problem for local owners since under present conditions this flooding will

Figure 6.5: Overwash deposits in cultivated lands in the Marquesa beach after the impact of the November 2001 storm. *Source*: Spanish Ministry of Environment.

occur with any significant Easterly wave storm impacting the Ebro delta coast (Jiménez et al., 2005).

Existing plans to cope with this situation are based on modifying the land use: adaptation to reduce vulnerability (see Chapter 1). Thus, Jiménez et al. (2000), in research for the Ministry of Environment and recently for the Generalitat de Catalunya (2004), proposed buying most of the affected land and incorporating it into the Natural Park. The proposal is to purchase from private owners a strip of about 107 ha of cultivated areas which are very close to the shoreline, and to incorporate this land into the public domain and to build a sand dune in the back of the beach to avoid hinterland inundations. The total estimated cost of such a scheme is about €4.8 millions.

Finally, we need to recognise that the affected area is located in an erosive coastal stretch with shoreline retreat rates greater than 3 m/yr (Jiménez & Sánchez-Arcilla, 1993). Owing to this we expect that the same coastal wave storms will have worse consequences in the near future (even under a steady wave climate) because the beach width acting as buffer for storm impacts will become progressively narrower. Thus, the proposed solution of creating a buffer as a natural area between the shoreline and the cultivated lands can be an example of adaptation to the system dynamics conditions.

The second agent we used to analyse the vulnerability of the cultivation function in the Ebro delta is RSLR, which has been already considered by Sánchez-Arcilla et al. (1998). In the long-term perspective, this is one of the most hazardous drivers affecting the delta, since the delta is a very low-lying environment with about 50% of its surface area lying below +0.5 m above mean sea level.

To obtain an estimate of the deltaic vulnerability to RSLR with regard to cultivation, we follow the approach of Sánchez-Arcilla et al. (1998) where the deltaic plain to be inundated under a given RSLR is estimated for three different scenarios: (i) *potential*, which is given by the area of the surface below the targeted sea level, (ii) *direct*, which is given by the area of the surface below the targeted sea level with direct connection to the sea and (iii) *storm-connected*, which also includes areas not directly connected but where the dune crest is easily breached during storm events.

Thus, for a RSLR of + 0.50 m (Sánchez-Arcilla et al., in press) values of the potential impact on the cultivation function of –0.42, –0.11 and –0.19 for each of the above-mentioned scenarios, respectively, are obtained. These values only consider direct inundation effects and they must be complemented with estimates of hydrological changes in the delta. These RSLR induced changes should vary the soil salinity across the deltaic plain and, thus, should affect the freshwater irrigation requirements needed to maintain the area's agricultural productivity. Again, these values are measured in terms of the ratio of the affected area to the total surface area dedicated to such a function in the delta. The large difference in the vulnerability estimates between the direct estimation and the others is due to the potential role of existing infrastructures — roads, channels and levees in the deltaic plain — in preventing the inundation of impounded low-lying areas.

Bearing in mind that the deltaic plain is no longer receiving sediment supplies from the Ebro river to compensate for sea level rise, together with the expectation of a climate-induced sea level acceleration, it is clear that this vulnerability analysis should be considered as realistic for the 21st century. Moreover, this analysis also indicates that to maintain the cultivation function in the Ebro delta into the distant future, means it will be necessary to have a proactive policy to compensate for the sea-level rise effects, or accept a move to polders such as in the Netherlands today.

Additionally, and regarding the economic implications of this agricultural deltaic function, it would be also necessary to consider, in any long-term analysis, the maintenance of subsidies which artificially keep the benefits of rice production at present high levels. Also, the role of rice fields in supplying food for the thousands of birds living or hibernating in the delta has to be considered if we are to analyse the full implications of this potential sea level rise effect. This illustrates one of the possible links between the cultivation function and the natural one.

Conclusions

At the core of this chapter lie the methodological and technical challenges of assessing deltaic vulnerability and sustainability. Vulnerability has been defined as the damage potential at a given timescale and for a predefined coastal sector. Sustainability, on the other hand, makes reference to the damage potential versus the recovery potential (natural or human induced) for a given time and space scale, where the human induced recovery is a function of the adaptive management policy pursued during the selected timescale.

Both concepts can be expressed in dimensional terms (with valuation units such as area or monetary value) or in dimensionless terms. In this later case the damage potential is quantified as a ratio of the actual damage or harm to that for a reference state. The concepts can consider the biophysical and social dimensions and, in consequence, they can be evaluated in any unit of measure able properly to represent/quantify the sub-system or system of interest. As an example, the geomorphic vulnerability of the system has been assessed in terms of sub-aerial surface changes. These can be obtained for a component of the coastal system (it would then gauge partial vulnerability or sustainability) or for the whole coastal system, and thus assess total vulnerability or sustainability.

The approach to be selected to quantify both concepts will depend on the level of knowledge available for the coastal section being considered. In this respect aggregated or bulk estimators such as the deterministic values used throughout this chapter can offer a number of advantages in supporting coastal management decisions in the context of major uncertainty or limitations of knowledge. Thus, they provide information on the present and future state of the deltaic system (or a given component). Moreover, the approach we present is able to discriminate between the different positive and negative contributions to such a state and, in consequence, it will help coastal managers to decide the type and the focus of corrective or adaptive actions when needed. Because of this, and regardless of the way they are quantified, both concepts should provide support in reaching more objective decisions in coastal zone management.

One of the direct consequences of applying these concepts to assess the state of a coastal deltaic system is that they offer a short cut for transmitting to coastal populations the inherent system dynamics and the finite carrying capacity of such ecosystems. In this chapter, this application has been undertaken in terms of fluxes or functionality. In the latter case, it has been applied to the two most common functions in deltas: nature conservation values and agriculture. The highest system vulnerability was for the natural component, although depending on the weighting scale used (which can be defined by the decision maker) different values were found.

Because of the current uncertainties and knowledge limits about the functioning of coastal systems, the simple approach we advocate nevertheless offers a number of advantages as a complement to detailed simulations of coastal dynamics. In this respect the bulk estimators of sustainability and vulnerability should be a natural complement or supplement to detailed simulations of coastal morphodynamics or ecosystem dynamics, which are as yet often not always robust enough to support sensible coastal management decisions.

Acknowledgements

This work has been partly undertaken within the framework of the FLOOD*site* project funded by the EU under contract GOCE-CT-2004-505420. The second author would like to thank the government of Catalonia for its support through the University Research Promotion Award for Young Researchers.

References

Baumann, R. H., Day, J. W., & Miller, C. A. (1984). Mississippi deltaic wetland survival: Sedimentation versus coastal submergence. *Science, 224*, 1093–1095.

Blaikie, P., Cannon, T., Davis, I., & Wisner, B. (1994). *At risk: Natural hazards, people's vulnerability and disasters*. London: Routledge.

Cardoch, L., Day, J. W., & Ibañez, C. (2002). Net primary productivity as an indicator of sustainability in the Ebro and Mississippi deltas. *Ecological Applications*, 12, 1044–1055.

Casanova, D. (1998). *Quantifying the effects of land conditions on rice growth. A case study in the Ebro delta (Spain) using remote sensing*. Unpublished doctoral thesis. Landbouwuniversiteit Wageningen.

Cencini, C. (1998). Physical processes and human activities in the evolution of the Po Delta, Italy. *Journal of Coastal Research*, 14, 774–793.

Chen, X., & Zong, Y. (1998). Coastal erosion along the Changjiang deltaic shoreline, China: History and prospective. *Estuarine, Coastal and Shelf Science*, 46, 733–742.

Costanza, R., D'Arge, R., de Groot, R., Farber, S., Grasso, M., Hannon, B., Limburg, K., Naeem, S., O'Neill, R., Paruelo, J., Raskin, R. G., Sutton, P., & van den Belt, M. (1997). The value of the world's ecosystem services and natural capital. *Nature*, 387, 253–256.

Day, J. W., Martin, J. F., Cardoch, L., & Templet, P. H. (1997). System functioning as a basis for sustainable management of deltaic ecosystems. *Coastal Management*, 25, 115–153.

de Groot, R. S. (2006). Function-analysis and valuation as a tool to assess land use conflicts in planning for sustainable, multi-functional landscapes. *Landscape and Urban Planning*, 75(1–2), 175–186.

Generalitat de Catalunya. (2004). *Programa de actuaciones urgentes en las zonas costeras del Delta del Ebro afectadas por los temporales*. Unpublished Report. Departament de Medi Ambient i Habitatge.

Gouldby, B., & Samuels, P. (2005). *Language of risk. project definitions*. FLOODsite Project Report T32-04-01, EU GOCE-CT. European Commission.

Guillén, J. (1992). *Dinámica y balance sedimentario en los ambientes fluvial y litoral del Delta del Ebro*. Unpublished doctoral thesis. Catalonia University of Technology, Barcelona.

Guillén, J., & Palanques, A. (1997). A historical perspective of the morphological evolution in the lower Ebro river. *Environmental Geology*, 30, 174–180.

Houghton, J. T., Ding, Y., Griggs, D. J., Noguer, M., van der Linden, P. J., Day, X., Maskell, K., & Johnson, C. A. (Eds). (2001). *Climate change 2001: The scientific basis*. Intergovernmental Panel on Climate Change. Cambridge: Cambridge University Press.

Jiménez, J. A. (1996). *Evolución costera en el Delta del Ebro. Un proceso a diferentes escalas de tiempo y espacio*. Unpublished doctoral thesis. Catalonia University of Technology, Barcelona.

Jiménez, J. A., Canicio, A., Sánchez-Arcilla, A., & Ibañez, C. (2000). *Caracterización de la problemática y alternativas de gestión y actuación para las costas del Delta del Ebro. Priorización de los problemas a corto plazo y soluciones*. Technical Report. Dirección General de Costas, Ministerio de Medio Ambiente.

Jiménez, J. A., & Sánchez-Arcilla, A. (1993). Medium-term coastal response at the Ebro delta, Spain. *Marine Geology*, 114, 105–118.

Jiménez, J. A., Sánchez-Arcilla, A., & Valdemoro, H. I. (2005). *Effects of storm impacts in the Ebro delta coast*. FLOODsite Project Report T26-05-10 EU GOCE-CT. European Commission.

Jiménez, J. A., Sánchez-Arcilla, A., Valdemoro, H. I., Gracia, V., & Nieto, F. (1997). Processes reshaping the Ebro delta. *Marine Geology*, 144, 59–79.

Kelly, P. M., & Adger, W. N. (2000). Theory and practice in assessing vulnerability to climate change and facilitating adaptation. *Climatic Change*, 47(4), 325–352.

Klein, R. J. T., & Nicholls, R. J. (1999). Assessment of coastal vulnerability to climate change. *Ambio*, 28(2), 182–187.

Menéndez, M., Hernández, O., & Comín, F. A. (2002). Spatial distribution and ecophysiological characteristics of macrophytes in a Mediterranean coastal lagoon. *Estuarine, Coastal and Shelf Science*, 55, 403–413.

Sánchez-Arcilla, A., & Jiménez, J. A. (1997). Physical impacts of climatic change on deltaic coastal systems (I): An approach. *Climatic Change, 35*, 71–93.

Sánchez-Arcilla, A., Jiménez, J. A., & Valdemoro, H. I. (1998). The Ebro delta: Morphodynamics and vulnerability. *Journal of Coastal Research, 14*(3), 754–772.

Sánchez-Arcilla, A., Jiménez, J. A., Valdemoro, H. I., & Gracia, V. (in press). Implications of climate change on Spanish Mediterranean low-lying coasts. The Ebro delta case. *Journal of Coastal Research*.

Stanley, D., & Warne, A. (1998). Nile delta in its destruction phase. *Journal of Coastal Research, 14*, 794–825.

Valdemoro, H. I., Sánchez-Arcilla, A., & Jiménez, J. A. (in press). Coastal dynamics and wetlands stability. The Ebro delta case. *Hydrobiologia*.

Woodroffe, C. D., Nicholls, R. J., Saito, Y., Chen, Z., & Goodbred, S. L. (2006). Landscape variability and the response of Asian megadeltas to environmental change. In: N. Harvey (Ed.), *Global change and integrated coastal management. The Asia-Pacific region, coastal systems and continental margins series*. Vol. 10. New York: Springer (in press).

Chapter 7

Local Communities under Threat: Managed Realignment at Corton Village, Suffolk

Sylvia Tunstall and Sue Tapsell

Introduction

This chapter explores the issue of social vulnerability to flooding and, more particularly, coastal erosion through a case study example of Corton village, on the east coast of England. It examines conflicts between community involvement, local community needs and preferences and the strategic and long-term principles of coastal zone management. Vulnerability is here defined very simply as the potential for a system (here a social system: individuals and communities) to experience harm with the focus on social and economic impacts from natural hazard events at the coast. Flooding and coastal erosion, the two main natural hazards affecting the UK coast, impact on social and economic life of communities in different ways (Table 7.1). Most importantly, erosion is effectively irreversible: once land has been eroded or the cliff-top has receded so that property cannot safely be used, then the land and property cannot be regained and this means that the future options for the use of that land or property are also lost. Flooding, in contrast, results in loss of capital stock in damaged buildings, stock and infrastructure which can be repaired or replaced in time, damage to the land itself through salt water flooding and disruption to businesses and lives for a period of time (Penning Rowsell et al., 2005).

While the tangible impacts measurable in money terms are the main focus of policy, strategies and project appraisal, research has indicated that residents in flood risk areas often see the intangible impacts, particularly the disruption to their lives and the stress and worry that flooding causes, as more significant impacts. Floods in the UK are mild by global standards and significant loss of life has been rare, excepting in Lynmouth in 1952 and on the East Coast in 1953. Furthermore, property is usually abandoned and access denied before lives can be put at risk in areas affected by erosion. However, studies indicate that flooding has a significant effect on the physical health and, more particularly, the mental health and well-being of flood victims in the longer term as well as the short term as people struggle to recover from a flood (Tunstall, Tapsell, Green, Floyd, & George, 2006). Results from quantitative and qualitative studies on the intangible impacts of flooding on households and individuals are summarised in Table 7.2.

Managing Coastal Vulnerability
Copyright © 2007 by Elsevier Ltd.
All rights of reproduction in any form reserved.
ISBN: 0-08-044703-1

Table 7.1: Social and econmic impacts of flooding and erosion on individuals and communities.

Form of impact Measurement	Social and economic impacts			
	Direct		Indirect	
	Tangible	Intangible	Tangible	Intangible
Flooding	• Damage to land • Damage to buildings • Damage to contents • Damage to infrastructure	• Loss of life • Physical health effects • Mental health and stress effects • Damage to heritage or nature conservation sites	• Disruption/loss of production and trade • Disruption to traffic • Disruption to/ loss of livelihoods	• Disruption to households during recovery • Disruption to community activities and life • Reduced confidence in area/planning blight • Disruption to social networks
Erosion	• Loss of land • Loss of buildings • Loss of contents • Loss of infrastructure	• Loss of life • Mental health and stress effects • Loss of heritage or nature conservation sites	• Loss of productive capacity and businesses • Loss of transport connections • Loss of livelihoods	• Loss of population through out-migration • Reduction in community activities and life, social capital • Reduction/loss of recreation and amenity value at the coast • Reduction in confidence in the area/planning blight

Table 7.2: Intangible impacts of riverine and coastal flooding an individual and communities.

Intangible impacts on households	Coastal flooding 1980s mean subjective rating[b]	River flooding: England and Wales[a] 2002 mean subjective rating[b]	Indicator of impact: River flooding England and Wales 2002[a]	Illustrative comments from qualitative studies of river flooding[c]
Disruption: time taken to get household back to normal	9.1	7.8	Mean time taken to get the house back to normal: 31 weeks	'It is like your life's taken away from you for weeks and weeks, you can not sit down, you can not read a book, you can not watch television, you have just got nothing, the house is just a building site'. Female, Todmorden, 2000.
Stress of the flood	8.0	7.1	33% reported shock symptoms as an immediate effect of the flood 55% reported anxiety when it rains	'Let me tell you, the shock it is to watch a river suddenly start coming towards your door and there is not a thing you can do and there is nowhere you can go'. Male, Todmorden, 2000. 'I felt like a robot. I cried but it was all more a dream. It was like, I just didn't have any emotions.... I just wanted to forget, we were all a bit traumatised' Male Todmorden, 2000.
Having to leave home	6.0	7.0	64% evacuated, some into caravans, for a mean of 23 weeks	'My family were on top of each other in this caravan, a 21-year-old daughter and 19-year-old son.... There's no privacy ... my son went absolutely off it the other night ... he said I am so fed up'. Female, South Church, 2000.

(Continued)

Table 7.2 (Continued)

Intangible impacts on households	Coastal flooding 1980s mean subjective rating[b]	River flooding: England and Wales[a] 2002 mean subjective rating[b]	Indicator of impact: River flooding 2002[a]	Illustrative comments from qualitative studies of river flooding England and Wales[c]
Worry about future flooding	4.0	6.6	27% very worried, 39% somewhat worried about flooding within the next 12 months. Worry depends upon whether or not flood defence scheme is promised/provided	'It takes your security away, your confidence in what should be a safe environment. Your home should be a safe comfortable haven for you but it isn't any more' Female, Banbury, 1999. 'The anxiety is still very much there … its just this awful sense of menacing foreboding that keeps happening everytime there's heavy rain' Female Banbury, 1999.
Loss of irreplaceable items	6.9	5.6	89% lost irreplaceable items or items of sentimental value	'My daughter was in tears at four o'clock in the morning trying to dry out old photographs of her gran.… That's the stuff that hurts most'. Male, Banbury, 1998. 'One day its your home … in our case we bought a house and it took us 25 years to make it a home, and its just gone'. Male West Auckland, 2000.

Health effects	5.0	4.5	Percentage of respondents reporting health effects: Immediate physical: 54%, longer-term physical: 33% Psychological: 72%. Percent currently scoring 4+ on general health questionnaire12, indicating mental health problems, flooded 25%, at risk 10%, Health Survey for England 1998 16%	'I don't know whether it was the shock of it or what. My voice went and then I got a terrible bad chest. I've got another bad chest now and I have never suffered with my chest in my life'. Female West Auckland, 2000. 'I mean I notice that … I just don't feel well as I normally would'. Female, Todmorden, 2000. 'The house is back to normal, but we are not back to normal in the mind' Male, Banbury, 1999. 'We just didn't think it would last this long. We are all upset but…. I'm sorry, I'm just so tired, I don't sleep, I wake up and there is water around the bed every night, I see water round the bed'. Female, Todmorden, 2000.
Overall seriousness of the effects of the flood on the household	7.6	7.3		
$N =$	42–135	656–973		

[a]Risk and Policy Analysts Ltd/Flood Hazard Research Centre (2004).
[b]Rating of the effects of flooding on the scale 0 = no effect, to 10 = extremely serious effect.
[c]Tapsell et al. (1999) and Tapsell and Tunstall (2000, 2001).

Much of this data on intangible impacts is for river flooding. However, the broad pattern of impacts appears to be the same for coastal and river flooding with some indications that impacts at the coast are more severe (Table 7.2).

Although research evidence is not available, it is likely that residents and businesses in areas affected by erosion suffer stress due to the uncertainty over their future there. Certainly, Members of Parliament, Councillors and others have set up organisations because of their concern about uncertainty over policy on erosion along England's east coast, for example, the Coastal Concern Action Group (www.happisburgh.org.uk). For flooding, the impact of direct damages can be mitigated by property and contents insurance that is available on the market for most properties. This insurance generally covers property holders against the risk of flooding and provides the main mechanism for compensating households and people with businesses affected by flooding in the UK. However, insurance does not cover the effects of erosion and, generally, no such compensation for losses is available to property owners affected by erosion from central government according to its guidance on the topic (www.defra.gov.uk/environ/fcd/policy/mrcomp).

The Corton case study illustrates, at the local level, the issues raised for such communities caught up in a process of government policy change and facing the prospect of managed realignment of their coast. In the face of accelerated sea-level rise and potential increased storminess at the coast, policy makers have recently come to recognise the need to work with natural processes through soft-engineering techniques and for 'managed retreat or realignment'. The case for this adaptive response to flooding and coastal erosion is set out in the government's latest strategy document 'Making Space for Water' (DEFRA, 2004).

'Managed retreat' was the term first used in England. However, this has fallen out of favour and has been replaced in official documentation by the term 'managed realignment' which does not have the negative connotations in relation to coastal management of 'retreat in the face of the enemy' rather than a rational management choice (Rupp & Nicholls, 2002). In this, actively managed defences are set back to a new line inland of the original or to rising ground or natural processes of erosion or flooding are allowed to occur in a managed way.

Most managed realignment schemes that have been implemented in England so far, including the pilot scheme in the Blackwater estuary in Essex, have been in rural areas and have had the objective of creating space for new intertidal habitats including tidal mudflats and salt marshes, subject to 'coastal squeeze'. In addition to providing valuable habitats, these schemes also act as 'natural' flood defences. However, it is now being argued that it is unsustainable in the long term to defend many small settlements along eroding coasts such as the Norfolk and Suffolk coasts because, when defended, they become promontories and deprive the coastline elsewhere of sediment supply and thus create problems there. Furthermore, such small settlements have too few assets to provide economic justification for defence schemes (North Norfolk District Council, 2004).

The terms, either managed retreat or managed realignment, are reproduced in this chapter as they were used at the time in relation to the case study. Managed realignment can be seen as an adaptation that reduces the vulnerability of the coastal system through allowing natural processes to work there. However, as we shall describe, it can also be seen as requiring a further adaptation on the part of individuals, community groups and organisations if their vulnerability is to be reduced. This needs careful analysis because it will inevitably affect the acceptability of such policies.

Corton Case Study Area and Research Methods

Corton is a cliff-top village on the eroding east coast of England, about three miles (2 km) north of the coastal town of Lowestoft, formerly a fishing port, now mainly a holiday resort with two beaches and about nine miles (6 km) south of the major Norfolk holiday resort of Great Yarmouth (Figure 7.1). At Corton, the partly vegetated cliffs of sandy boulder clay, sand and gravels are subject to erosion and slippage due to groundwater penetration. The southern part of the Corton Cliffs is a designated geological Site of Special Scientific Interest and continuing exposure of this part of the cliffs is required to maintain their geological interest. The cliffs were protected by 1960s defences comprising a 1.5-m wide concrete seawall

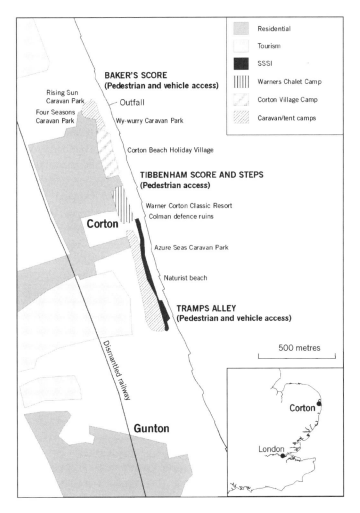

Figure 7.1: Map of the Corton seafront showing the location of the threatened holiday facilities and access to the seafront.

promenade and revetments and by a field of now decaying steel-piled timber groynes. Some of the defences were renewed or rebuilt in the 1980s. Further work on the defences was carried out in 2002–2003 and subsequently. At present, the beach is variable in height and low beach levels only allow tide restricted access to limited sections of the beach.

The village has about 1000 residents, a church, two public houses, a school, a post office and village shop. Its origins as the summer residence of the nineteenth century 'mustard magnate', Jeremiah J. Colman, give the village a special character. A benefactor of the village, in the 1870s–1890s, he built coastguards houses and other cottages and the village hall, all with distinctive decorative local brickwork along the main street which runs parallel to the coast.

Corton has a long history both of erosion and as a holiday village. Several houses are known to have collapsed into the sea in the past. Although Jeremiah Colman put up seawall defences, remnants of which are visible in the sea today, to protect his cliff top estate, 'the Clyffe', the property had to be abandoned and the cliff-top grounds were sold off to become the Corton Pleasure Gardens in 1917. This, in turn, became in 1924 a holiday camp for the Civil Service Holiday Association, which was taken over as Warner Holiday Camps Ltd in 1946. From the 1940s, other holiday and caravan camps have been established along the cliff-tops. Today, the Corton seafront caters for staying visitors, providing mainly affordable holidays for low-income families and older people in holiday villages and caravans, as well as for local residents (Table 7.3). There are few day visitors. The cliff-top holiday facilities are the assets most immediately vulnerable to erosion (Figures 7.2 and 7.3) but in the longer term, the main street behind them and access road to the village, the school, and many other properties would be threatened.

Middlesex University was asked by Waveney District Council, the local authority responsible for the Corton coast, to undertake a preliminary investigation to examine public preferences for, and the likely recreation benefits of, the various options for protecting Corton's seafront. This involved site visits, semi-structured interviews with 13 key informants such as managers of holiday facilities, with 12 local resident and 52 questionnaires completed by visitors staying at three local sites in the summer of 2000 (Tunstall & Penning-Rowsell, 2000). In the following year, we were also commissioned to carry out a contingent valuation survey of Corton residents and visitors using a structured questionnaire to assess the use they made of the Corton seafront and the value they would place on their recreation there under various adaptation options for coastal management and protection. Interviews were completed with 163 residents in about a third of the households in the village. In addition, 304 holidaymakers at five sites were interviewed (Table 7.3) (Tunstall, 2001).

Policy Change and Uncertainty in Corton

The Corton case study illustrates the process of government policy transition and uncertainty over erosion rates. In response to central government's urging a more integrated and co-ordinated approach to coastal protection, Waveney District Council, the authority responsible for Corton, issued a consultation document 'Planning for the Coast' as an initial step, setting out suggested planning policies for areas at risk from erosion prior to a

Table 7.3: Holiday facilities at Corton.

Name of facility	No and type of units	Maximum number of bedspaces	Type of visitor catered for	Estimated total adult visits to seafront per annum	Number of interviews completed: summer 2001
Warner classic resort (one of a number of holiday resorts owned by Warners)	205 chalet style units	385	Adults only; mainly older people; those on low income; affordable week holidays with entertainments provided; open year round	65,000	166
Corton Beach Holiday Village (privately owned)	124 beach villas (21 privately owned)	694	Affordable family beach holidays; open April to end October	22,000	105
Azure Seas Caravan Park (one of two caravan sites owned by the company)	70 static caravans (privately owned) space for 30 touring caravans plus tents	About 600	Long term owners; open Easter to end October	19,000	19
Rising Sun Caravan Park (privately owned by family with long term Corton connections)	22 privately owned 6 berth static caravans	132 maximum	Many local long-term owners; open Easter to early January	NA	10
Wy-Wurry (privately owned by site resident)	11 static caravans; 6 privately owned	66 maximum	Adults only; open Easter to early January	NA	4
Four Seasons Caravan Park (privately owned by site resident)	3 static caravans for letting	12 maximum	April to early January	NA	–

NA=Not available.

Figure 7.2: The small 'Wy Wurry' caravan park ironically the most immediately vulnerable of Corton's holiday facilities.

Figure 7.3: Caravans located at the cliff edge in another Corton caravan park threatened by coastal erosion.

more comprehensive treatment of the coast through a shoreline management strategy (Waveney District Council, 1993). This document noted that although the toe and front of the Corton cliffs were secured by a sea wall and groyne system, the cliffs were rendered instable by the action of groundwater on the cliff face. However, tentative 75-year estimates of the cliff line after collapse had allowed recession to a safe angle were relatively reassuring for cliff-top businesses (apart from the Azure Seas and Wy-Worry Caravan parks). These suggested that the cliffs would stabilise before erosion reached most built property although land would be lost.

In line with the government's policy for a more strategic approach to coastal management and defence (for sediment cells or sub-cells), a new document that covered an extensive stretch of East Anglian coast, the Sheringham to Lowestoft Shoreline Management Plan (SMP) (Halcrow, 1996), was produced. It proposed that the policy for most of the Corton seafront should continue to be one of the 'hold the line' through maintaining the coastal defences, although the policies for areas immediately to the north and south of the village were for 'managed retreat' and 'do nothing', respectively.

However, a rather different view of the rate of erosion and of the appropriate policy response was taken in a document issued three years later as part of the more detailed Gorleston to Lowestoft Coastal Strategy Study (Halcrow Maritime, 1999). This indicated that if nothing were done to the coastal defences, five of the cliff-top holiday businesses and 50 residential properties would lie within the 50-year cliff recession line and that the businesses would become unviable after 10–17 years. The strategy also took the view that it was not sustainable in the long term to protect Corton while allowing erosion to proceed to the north and south of the village: as a hardened point, it would become a promontory. Furthermore, on initial review, there were insufficient economic benefits available to justify major coastal defence works there. Therefore, a changed policy of 'managed retreat' and progressive removal of defences to allow a natural coastline was recommended.

In the event, the more immediate vulnerability of the Corton seafront was graphically illustrated when the defences, the sea wall and revetments collapsed in three places due to undermining during storms over the winter of 2000–2001 (Figure 7.4). As a result most of the Corton beach and the access point in the centre of the village were closed for safety

Figure 7.4: Corton's coastal defences, sea wall, promenade and revetments collapsed due to undermining in the winter of 2000–2001.

reasons for part of the holiday season. Access was only possible to small sections of the beach at the extreme ends of the village and it was no longer possible to make the circular walk along the full length of the Corton seafront and back through the village as residents and visitors were accustomed to do.

Following on from the strategy, Waveney District Council with its consultants developed a range of scheme options, which represented different ways of adapting to erosion at the Corton coast.

Preferences for the Options

In the Middlesex University survey, for comparison with the options, residents and visitors were shown a drawing representing the Corton seafront in the condition it was in at the time of the interview and a brief description of the seafront in its current state was read out. The seafront was already badly affected by erosion and storm damage prior to the interview. Thus, the description included the following key points:

- The coastal defences and seawall have already collapsed in three places.
- There is no access to the beach from the centre of the village although this may become available.
- The only access point to the beach is at the extreme south edge of the village and most of the beach is closed.
- The beach is low and there is little beach at high tide.
- At times, there is a big drop down to the beach from the seawall walkway.
- The groynes are dilapidated.

Not surprisingly, interviewees had a poor opinion of the current condition of their seafront. Local residents, a majority of whom had experience of the Corton seafront going back 20 or more years, were consistently more critical than visitors. The proportions of residents rating the Corton seafront as very poor for walking, as a place to take children, for its beach and overall as a place to visit were 71, 58, 49 and 56%, respectively. For staying visitors the comparable proportions were 45, 36, 18 and 27%. About half the staying visitors were making a first visit to the place. A majority of those who had been before noted a deterioration. Most local residents had earlier memories of Corton; some remembered a 'big beach' there from their childhood and almost all saw the Corton seafront as having changed markedly for the worse over time.

When respondents had considered the current situation, four options for adaptation at the Corton seafront were offered to them through drawings representing the seafront with the option in place and in descriptions that were read out to them. The wording of these statements, presented slightly amended in Table 7.4, was agreed with Waveney District Council and their consultants, who also provided the accompanying drawings. Table 7.5 summarises survey responses to the different proposed adaptations.

Respondents were told that if nothing were done to protect the cliffs at Corton, the seafront would change and after 10 years it would have the characteristics described in the 'do nothing' option. Around three-quarters considered that a visit would be much less

Table 7.4: Descriptions of the options for adaptation.

Do nothing	Hold the line for a limited period	Hold the line for a longer period	Managed retreat
• The cliff protection, sea well and walkway on top of the seawall have collapsed along almost the whole Corton seafront • There is no access along most of the beach because of hazards from seawall debris there • The only access to the beach is at the southern end of the village and to the beach there • The cliffs are eroding and changing to a more natural appearance and some clifftop visitor facilities are deteriorating with the prospect of closure • The groynes have almost decayed away	• The coastal defences have collapsed along large parts of the seafront and the cliff face is protected by rocks and possibly rock filled wire baskets • It is no longer possible to walk all the way along the foot of the cliff as only a few sections of the walkway remain in place • There is likely to be only limited access to parts of the beach in the centre of the village and to the beach at the southern end of the village • The beach itself is low and partly covered in rocks protecting the base of the cliffs. There is an narrow beach at high tide • The groynes are very dilapidated	• Where the cliff protection has collapsed, the cliff face is protected by rocks • There is a concrete slab walkway along the base of the cliffs. It is possible to walk all the way along the Corton seafront • Concrete steps lead from the walkway onto the beach • There is full access to the beach and along the seafront from three access points • There are new (rock) groynes along the seafront	• The cliffs faced with earth and vegetation have natural appearance • There are no coastal defences (no groynes) and no walkway at the base of the cliffs • There is no direct access to the beach from Corton village itself but there is access all long the beach from two extreme end points • The beach itself is narrow but probably walkable at high tide • The cliff is eroding and some visitor facilities are dilapidated with prospect of closure

Table 7.5: Public response to adaptations to erosion through scheme options.

	Do nothing	Hold the line for a limited period	Hold the line for a longer period — 50 years or more	Managed retreat
Enjoyment with option (%)	(%)	(%)	(%)	(%)
Much less	73	17	2	51
Less	13	23	*	21
Same	11	25	1	13
More	3	27	10	9
Much more	*	8	87	5
Percentage visiting elsewhere with the option	57	33	0	45
Preference for option (%)				
Strongly against	85	18	*	58
Against	12	41	2	23
Neither	1	18	2	9
In favour	2	22	12	6
Strongly in favour	*	2	84	4
Number of respondents	464	461	461	457

* <0.05%

enjoyable under this option. This option was strongly condemned by residents and visitors alike and was the least popular option. There was a strong expectations that the authorities should 'do something' to protect the coast.

A majority of those surveyed would choose to make visits elsewhere were nothing done. For residents and visitors with cars, there were alternative beach sites within a short drive in Lowestoft, Gorleston, Great Yarmouth and in other places, some of which offered better beaches and facilities. However, getting the car out and travelling even short distances in holiday traffic were viewed as less convenient than walking to the local beach. The closure of the local beach had certainly disappointed some of the visitors staying in the holiday camps. The main reasons for being opposed to the 'do nothing' option were that it would make the beach unusable and inaccessible, cited by 21%, dangerous because of the decaying defences mentioned by 21% and/or unsightly or ugly, cited by 18%. The comments of those interviewed on this option further illustrate the nature of their opposition.

> 'Looks dangerous, dirty, no access for disabled, still dodgy for able bodied people. Appalling that we are in the 21st century and we cannot do anything about this.'

> 'You cannot walk along, this is my pleasure when I come to a resort, the attraction of a coastal visit is to walk along the sea edge.
> Want to be able to visit with grandchildren and watch nature, won't be safe, people would lose their homes and have limited access, won't bring the holiday trade into town.'

The main view of the few respondents who favoured the 'do nothing' option was that the deterioration had gone too far and that, therefore, it was best to let nature take its course.

Although the 'hold the line for a limited period' option was seen as some improvement on the current state, there was very little enthusiasm for this option (Table 7.5). A key concern was the lack of access to the seafront, mentioned as a reason by 27% of those against this option. It was not favoured by 17% because it was considered to be a partial, short-term measure and a poor investment. It is interesting that even older people took a long-term view and wanted a long-term solution. There were fears, too, that the rocks would prove dangerous and the scheme unsightly, mentioned by 8 and 9%, respectively. It was, thus, seen as addressing the need to protect the coast but not the recreational needs of visitors. There were critical comments about the rock protection for the cliffs with this option.

> 'Looks ugly, would cost a great deal of money and would be better spending more to make a good job of it.'
> 'If that is all we are going to get, we will take it ... But not a long-term problem solver. Might collect rubbish in the stones (rock boulder protection).'
> 'It would strengthen the cliff but does not look a nice place to go. You can't walk a long way.'

The 'hold the line for a longer-term period — 50 years or more' was the option that residents and visitors wanted and expected from the authorities. While the policies of the government for the coast have moved on, attitudes and expectations of the public, on the evidence of Corton, have still to shift. Generally, this option restored and maintained the seafront and the facilities there as they had been before the recent collapses and in earlier times. Respondents were fulsome in their praise for this most favoured option. Clearly, the key perceived advantages of this option over the others were the much greater access from the village and along the beach on a walkway and onto the beach via steps mentioned by 29% and the longer-term nature of the solution. This option was also favoured as nice looking and natural by 13% despite its highly engineered character incorporating concrete facings to the cliffs, rock protection and rock groynes. Safety for beach users (11%) and protection for homes and businesses (11%) were among other reasons offered for supporting this option. The following comments illustrate what residents and visitors liked about this option.

> 'Full access for everyone including the elderly, looks nice, long term, safe.'
> 'Neat and tidy, good access to the sea, strong groynes, can take push chairs down to the beach.'
> 'Brilliant. Plenty of beach, plenty of walkway. Easy access, very pleasing to the eye.'

Barriers to Managed Realignment at Corton and Elsewhere

Corton provides an example of managed retreat recommended for a developed coast in England and of the attitudes towards this form of adaptation of a local community. Other small villages and settlements along the East Anglian coast are facing the same prospect (see North Norfolk District Council, 2004). In the survey, respondents were told that this retreat option would allow the Corton coast to go back to a more natural condition. Collapsed defences would be removed leaving a natural eroding cliff face and beach leaving the seafront after 10 years as described in Table 7.4. Presented in this way, managed retreat was almost as unpopular as the 'do nothing' option (Table 7.5).

Why were people, particularly local residents, so strongly opposed to this option and what can the Corton example tell us of likely responses elsewhere?

First, the policy for managed realignment in Corton and for settlements along the east coast of England and elsewhere must be viewed in its historic context. The east coast flooding of 1953, the worst coastal flooding in England in the last century, resulted in changes to coastal management policy that endured almost into the twenty-first century. The impact of this event was huge: approximately 24,000 houses were flooded, 32,000 people were evacuated and over 300 people died in the flooding. Although Corton itself suffered no loss of life, community memory of the event and of the deadly power of the sea remains vivid along the east coast.

The flood event acted as a catalyst for policy change. Two long-term adaptations followed from the event. A new permanent warning system along the east coast was initiated. Since its introduction in 1953, this service and warnings in general have evolved as a key priority element in flood risk-management strategy and as an adaptation that can make communities and individuals more resilient in the case of flooding. More relevant to the Corton case was the decision that the 1953 flood level should be the maximum level of protection provided except where particularly valuable property was at risk. As a result defences that have lasted to the present day were built along the English coast (Johnson, Tunstall, & Penning-Rowsell, 2004). This long-term policy focus on structural flood defences has, perhaps, built up a public expectation of, and support for, such interventions to protect the coast both from flooding and erosion.

Second, the realisation even among policy makers nationally that this policy was no longer viable and the shift from a policy of flood defence to one of the flood risk management (Environment Agency, 2003) is of relatively recent origin. As indicated above, the policy of managed retreat at Corton was also recent, dating from 1999. From the 1980s, there had been a government policy change away from protecting agricultural land from flooding and erosion reflecting the gradual decline in importance attached to the agricultural sector and to self-sufficiency in food production in the economy since the Second World War.

From the 1990s, there has been a gradual transition towards a flood risk-management approach. During this period the government's adoption of sustainable development as a guiding principle for policy meant that social, economic and particularly environmental issues had to be taken into account in river, coast and land management. EU legislation, e.g. the Habitats Directive 1992 was also influential. At the same time growth in knowledge and understanding of climate change and of coastal processes and the development of 'soft' and bio-engineering techniques encouraged approaches to flood defence that sought to work 'with' nature rather than 'against' it (Tunstall, Johnson, & Penning-Rowsell, 2004). Given the recency of the national policy change, and the fact that it was based on scientific understanding of the impacts of climate change, of coastal processes and bio-engineering techniques, it was not surprising that public appreciation and support for the approach lagged behind that of decision makers.

At the time of the Middlesex University preliminary investigation in the summer of 2000, a year after the publication of the proposal for managed realignment at Corton, it was clear that awareness of the proposals was low and a major adjustment in thinking would be required for the owners and managers of the cliff-top businesses to accept the idea that the Corton seafront was no longer to be protected. At this stage, many found it inconceivable that the government would countenance abandonment and one holiday camp owner had made plans for a major investment to refurbish his property. Among businesses, as well as residents, there was a reluctance to contemplate radical change. In the survey, too, both residents and visitors strongly disagreed with the idea that it was best to let nature take its course at the coast and to allow erosion to continue and with the proposition that it was not worth spending money to preserve a beach because people could always go somewhere else.

Middlesex University survey data from 1987 and from Corton indicate that residents continue to attach importance to protecting most types of land use at the coast, including farmland no longer given priority by government (Table 7.6). Not surprisingly, the highest priority for residents was for protecting houses. There was some indication in the 1987 survey data of a decline in the priority given to protecting land uses at the coast with distance from the coast.

Third, there is evidence that residents and visitors in England are generally conservative in their attitudes towards the coast. We found in the preliminary investigation that some residents had long connections with the village and the seafront and radical change that would obliterate historical and family links with the past was rejected by them. People felt that it would be a shame to lose what had been part of the village for so long. This desire to preserve the coast as it had been in the past and reluctance to accept the workings of natural processes at the coast has been found to be a common feature of the public's perceptions of English coastal sites (Tunstall & Penning-Rowsell, 1998) and is reflected in these quotations from Corton residents and business owners in the preliminary investigation.

> 'You want things to stay the same forever.'
> 'A nice beach like we had years ago.'

Fourth, research shows that people come to accept and indeed to value forms of intervention, such as seawalls with promenades and groynes at the coast. Such facilities can be an important element in the recreational experience at coastal resorts particularly, places attracting elderly people (Tunstall & Penning-Rowsell, 1998). Certainly at Corton, a walkway on top of a seawall was regarded as an additional amenity. Although, the natural look of the beach and its greater potential for wildlife under the managed realignment option had some appeal, this was not sufficient compensation for the lack of access to the beach from the village, lack of a promenade walk along the base of the cliffs and the impact on

Table 7.6: Priority for protecting different types of use of the coast.

Land use	At coast[b]		Mean rating of priority[a]					
			30 minutes from coast[c]		One hour from coast[c]			One and a half hours from coast[c]
	Corton residents	Corton visitors	Norwich residents	Hull residents	Hendon residents	York residents		Cambridge residents
Houses	5.8	5.5	4.6	4.4	4.2	3.7		3.4
Beaches	5.3	5.3	4.6	4.0	3.3	3.3		3.9
Farmland	4.7	4.8	4.2	4.4	3.9	3.4		3.3
Bird/Nature reserves	4.7	5.1	4.6	3.6	3.7	3.3		4.1
Public open space	4.6	4.8	4.0	3.5	3.5	3.2		3.0
Promenades	4.6	4.9	3.5	2.8	1.7	1.8		1.1
Heritage/archaeological sites	4.4	4.7	4.3	3.2	2.8	2.4		2.8
Holiday facilities e.g. chalets	5.0	5.1	NA	NA	NA	NA		NA

[a]Scale: 6+ highest priority to 0 = lowest priority.
[b]Survey date 2001.
[c]Survey date 1987.
NA=Not available.

the holiday facilities and the village in general. In Corton, and it would probably be the case elsewhere, loss of access was a key factor in opposition to managed realignment, cited by 17% in Corton. With the cliffs allowed to erode, it would not be possible to fix access steps down the cliff face from the centre of the village.

Furthermore, natural eroding cliffs were regarded as a source of danger from falling cliff material and a reason for opposing managed realignment, cited by 16%. Restoring the seafront to a more natural condition found favour with only about 10% of the survey respondents, who liked the natural appearance. However, even this minority had reservations on account of the lack of access and safety concerns. The following comments on the managed retreat option by Corton residents and visitors illustrate these attitudes.

> 'Not for elderly people. A lot of people can't walk on sand and need a pathway.'
> 'Although it looks lovely and reminds me of my childhood, we would have no access to the beach at low tide, our families would not be safe visiting and the holiday trade would close down.'

There are also indications that 'natural' coastal sites attract a different kind of visitor in search of a different experience from those who go to developed sites (Penning Rowsell, Coker, N'Jai, Parker, & Tunstall, 1989). Thus, the transformation of a site from one with human constructions to a 'natural' site is unlikely to meet the requirements of current users. However, in the future, new users for whom the natural aspect of the site had a special appeal might be attracted. Managed retreat was the only option on which there was a divergence of opinion between local residents and staying visitors, with the locals more hostile to this form of adaptation than the visitors (74% of residents compared with 50% of visitors strongly opposed). This suggests that those who have a greater interest, and possibly a stake in a site, find radical change there more difficult to contemplate. Future residents may, of course, see things differently. It is possible too that, with time, current residents would come to accept a coast returned to a more natural state.

Fifth, in the UK to be eligible for central government grant aid, coastal protection schemes have to be justified in terms of national economic benefits. Losses to the local tourist trade and economy that can be taken up elsewhere within the UK will not constitute national economic losses and will not be taken into account. Local authorities and local people are, however, concerned about local sustainability in which local social, economic and environmental factors are balanced. Local economic issues are of particular concern to local authorities and local communities in areas where unemployment is relatively high and incomes relatively low, as is the case in Waveney District in which Corton lies.

The impact and adaptive capacity varied among the businesses and their customers. In principle, owners could move their caravans to new sites. However, those interviewed in the preliminary investigation valued their cliff-top sites and views (Figure 7.3) and doubted whether, given planning constraints on caravan sites, comparable alternatives could be found. Holidaymakers could easily find alternative places for their holidays. The holiday facilities that were part of a holiday group could transfer their Corton business to other sites. However, some small caravan site owners had invested their life savings in their businesses and also lived on the site: they would lose their homes as well as their incomes in time if the coast were not protected. There was considerable community sympathy for these small businesses, which lacked the resources of the larger enterprises to relocate and start again.

Local people in Corton did not have the negative feelings towards the visitors that seaside residents sometimes feel about holidaymakers, perhaps because Corton attracted mainly older people in search of a quiet holiday and because Corton had a long history as a holiday village. Indeed, many Corton residents were aware of the benefits that the holiday trade and visitors brought to their community not only in terms of local jobs but also in terms of support for local facilities: the post office, shop, pubs and even the church fete and jumble sales. Residents also occasionally made use of the facilities at the holiday camps such as entertainments and swimming pools there. None of those interviewed relished the idea of losing the holiday camps and caravan sites along the cliff-tops to the sea, although these developments could be regarded as disfiguring the landscape. Many thought that their loss would have a devastating effect on their community

> 'No heart to the community, traditional professions and lifestyles would die, the village would go, the last remaining old properties would go into the sea.'
> 'If they (holidaymakers) were to go, the whole village would die.'

Local people, too, felt a certain attachment to their village with its nineteenth century coastguards' cottage, village hall, which would in the long term be threatened by erosion and sympathy for the residents whose properties would be lost. Other village facilities, a shop, a village school, public house and the Methodist church would eventually be lost under a managed realignment option and alternatives would have to be provided inland. Thus, the impacts and adaptations required were at a community level as well as at the level of the individual resident or business.

Overcoming Barriers to Managed Realignment at Corton and Elsewhere

The Corton study shows that a move from a 'hold the line' to a 'retreat the line' policy can be a shock to a local community, its residents and businesses and that policy makers need to consider very carefully how to introduce and carry out such a change. It also reveals significant barriers to the implementation of managed realignment on a developed coast in terms of public attitudes and expectations. It also provided some clues as to how these barriers may be overcome.

The case study highlighted the importance of time factors, in particular, the conflict between the long-term view of adaptation required for working with natural coastal processes and the short-term considerations affecting local authorities and their communities. Certainly, the survey showed that people need time to adapt psychologically and practically to physical and policy changes at the coast. The consultation draft of the Shoreline Management Plan (SMP) for the area (North Norfolk District Council, 2004) recognised this need.

The preferred long-term plan (over 100 years) for Corton remained to allow the cliffs to retreat and a more natural shoreline position to be attained so that Corton was no longer in a prominent position on the coastline. However, a short-term policy to hold the line to protect for up to 20 years from 2005 was also proposed in the SMP to give people the opportunity to make the major adjustments to their lives and businesses that will be required thereafter. The SMP anticipated that wall failures would occur within 20–30 years and in order to comply with the long-term plan, defences would not be replaced should they fail earlier than that.

The preferred medium-term plan (covering a time period from 2025 to 2055) was to allow cliff erosion to take place through a policy of managed realignment and to carry out only minimal interventions to manage and slow the retreat. It was estimated that there will be a loss of cliff-top assets of up to 40 properties under this policy and it will be a requirement to put in place prior to this time measures that will enable appropriate relocation of people and properties and facilities. In the longer term to the year 2105, overall, it is anticipated that over 100 properties will be lost as the cliffs retreat including the cliff-top holiday camps, the village school, the Methodist church, the village hall and a public house as well as the main road through the village.

Waveney District Council sought to develop a preferred short-term option scheme of 20 years duration that would satisfy as many of the amenity and recreational requirements of residents and visitors, particularly for access, as possible. Local people and community stakeholders were consulted on this option in December 2001 (Patterson, Glennerster, & Millar, 2004). Complex works implementing this preferred option scheme were carried out along the Corton seafront in 2002–2003. The scheme involved new timber access steps and renewed revetments and seawall with a narrow promenade and handrail along the foot of the cliffs all protected by massive Norwegian rocks at the toe of the cliffs (Figure 7.5).

A key lesson from the Corton case study concerns the importance actively engaging with the public on major changes at the coast. It has come to be generally accepted that 'interested parties' should, at minimum, be consulted to ensure that they are aware of the policy developments and are given the opportunity to express their views and concerns regarding coastal protection and sea defence policy either formally or informally (DEFRA, 2001).

Figure 7.5: Corton's coastal defences at high tide renewed in 2002–2003 with a narrow walkway on the sea wall and extensive rock cliff protection.

More recent guidance on shoreline management plans supports greater involvement of key stakeholders in the planning process. For the Gorleston to Lowestoft Coastal Strategy Study 1999, a conventional consultation exercise was mounted: an initial consultation letter seeking information and views was sent to a number of national and local organisations including local businesses and statutory authorities.

Subsequently, the local authority engaged very actively and widely with local businesses, residents and the parish council (the lowest tier of local government). Although there is no systematic research evidence to support this view, Waveney District Council believes that its success in engendering acceptance of the short-term preferred option and of the long-term policy of not holding the line was due to its high level of consultation with locally elected councillors and the local community (Patterson et al., 2004). Certainly, the fact that the local authority project manager was in post throughout the process and was observed to be a frequent presence in the village assisted communications. The local authority sent out newsletters and together with consulting engineers held public meetings and exhibitions and one-to-one discussions with individual residents and businesses and with the parish council throughout the scheme option development and implementation. This provided opportunities for information exchange and raised awareness and understanding of the constraints within which coastal management policies are set and delivered (Patterson et al., 2004).

With the short-term scheme in place, no campaign in opposition to the long-term managed realignment has developed in Corton whereas local MPs and others elsewhere along the coast have formed Suffolk coast authorities against realignment (SCAR). This organisation is calling for those who acquired property at risk from erosion at a time when the policy for their coast was one of protection to be compensated when the policy is changed. Furthermore, it does not appear that Corton properties threatened in the long term have been blighted since two of the holiday facilities have changed ownership and one has received substantial investment and two residential properties within the 50-year erosion zone have been sold successfully. Indeed, it has been suggested that the Corton approach of providing a short-term scheme to allow communities time to adjust to long-term managed realignment could provide a model for such policies on developed coasts.

What is less clear is whether nature will allow the community the time to adapt and whether the short-term defences will last for the 20 years intended. A severe coastal storm in December 2003 once again damaged the new defences resulting in the closure of some access points. However, additional funding was received for repairs and the promenade and access points from the village were re-opened and in use from mid-2005 and the defences remained intact at the start of 2006. Other uncertainties attaching to this approach involve the issue as to how and when the decommissioning of the short-term defences including the large quantities of rock should be achieved. These will clearly be politically difficult decisions for the local authority to take.

Conclusion

We can now see that managed realignment is a form of adaptation that can indeed increase the vulnerability of communities and is complex in that it requires a further significant adaptation on the part of community members: they have to relocate in the long term. This can engender antagonism within communities towards the policy and make policy implementation particularly difficult. Engaging with communities and local stakeholders as policy is developed and allowing time for communities to adjust to a change in policy may help, as Corton indicates. But of course, this phenomenon is highly geographically specific and we do not know how representative Corton is of similar situations on other eroding coasts. It is these differences that make the vulnerability of this type of coastal community difficult to predict and also make a strategic approach to vulnerability reduction difficult both to devise and to implement.

Acknowledgements

Waveney District Council provided the funding for the investigations and surveys. We are grateful to Waveney District Council staff, in particular, Mr Paul Patterson for organising site visits and providing information and to the residents, visitors and other stakeholders who participated in the research.

References

DEFRA (2001). *Shoreline management plans: A guide for coastal defence authorities*. London: Department for the Environment, Food and Rural Affairs (DEFRA).

DEFRA (2004). *Making space for water: Developing a new Government strategy for flood and coastal erosion risk management in England*. A consultation exercise. London: Department for the Environment, Food and Rural Affairs (DEFRA).

Environment Agency. (2003). *Strategy for flood risk management 2003–2008*. Bristol: Environment Agency.

Halcrow. (1996). *Sheringham to Lowestoft shoreline management plan sediment sub-cell 3b document*. Shoreline management plan strategy. Report for North Norfolk District Council, Great Yarmouth Borough Council, Waveney District Council and National Rivers Authority. Swindon: Halcrow.

Halcrow Maritime. (1999). *Gorleston to Lowestoft coastal strategy study*. Summary report and action plan consultation draft. Report for Great Yarmouth District Council and Waveney District Council. Swindon: Halcrow.

Johnson, C., Tunstall, S., & Penning-Rowsell, E. (2004). *Crises as catalysts for adaptation: Human response to major floods*. Enfield: Flood Hazard Research Centre.

North Norfolk District Council. (2004). *Kelling to Lowestoft Ness shoreline management plan*. SMP Document for Consultation. Cromer: North Norfolk District Council.

Patterson, P., Glennerster, M., & Millar, G. (2004). Corton coast protection. *Proceedings of the 39th Defra conference of river and coastal engineers, July 2004, Keele University*.

Penning Rowsell, E., Coker, A. M., N'Jai, A., Parker, D. J., & Tunstall, S. (1989). Scheme worthwhileness. In: Institution of Civil Engineers (Eds), *Coastal management*. London: Thomas Telford.

Penning Rowsell, E., Johnson, C., Tunstall, S., Tapsell, S., Morris, J., Chatterton, J., & Green, C. (2005). *The benefits of flood and coastal risk management: A manual of assessment techniques*. Enfield: Middlesex University Press.

Risk and Policy Analysts Ltd/Flood Hazard Research Centre (2004). *The appraisal of human-related intangible impacts of flooding*. Report to Defra/Environment Agency, R&D Project FD2005, Defra/Environment Agency.

Rupp, S., & Nicholls, R. J. (2002). Delft hydraulics. In: van Kappel, R. (Ed.), *Managed realignment of coastal flood defences: A comparison between England and Germany.* Prepared for 'Dealing with Flood Risk': An interdisciplinary seminar of the regional implications of modern flood management, 4 March 2002, Delft.

Tapsell, S.M. & Tunstall, S.M. (2000). *The health and social effects of the June 2000 flooding in the north east region.* Report to the Environment Agency, Thames Region, Enfield: Flood Hazard Research Centre.

Tapsell, S.M. & Tunstall, S.M. (2001). *Follow-up study of health effects of the 1998 easter flooding in Banbury and Kidlington.* Report to the Environment Agency, Enfield: Flood Hazard Research Centre.

Tapsell, S.M., Tunstall, S.M., Penning-Rowsell, E.C. & Handmer, J.W. (1999). *The health effects of the 1998 easter flooding in Banbury and Kidlington.* Report to the Environment Agency, Thames Region, Enfield: Flood Hazard Research Centre.

Tunstall, S. (2001). *Corton village coastal protection recreation survey.* Draft Final Report. Enfield: Flood Hazard Research Centre.

Tunstall, S., & Penning-Rowsell, E. (1998). The English Beach: Experiences and values. *The Geographical Journal, 164*(3), 319–332.

Tunstall, S., & Penning-Rowsell, E. (2000). *Corton village coast protection works: Preliminary investigation.* Enfield: Flood Hazard Research Centre.

Tunstall, S., Tapsell, S., Green, C., Floyd, P., & George, C. (2006). The health effects of flooding: Social research results for England and Wales. *Journal of Water and Health, 4*, 365–380.

Tunstall, S. M., Johnson, C., & Penning-Rowsell, E. (2004). Flood hazard management in England and Wales: From land drainage to flood risk management. *Proceedings of the World Congress on Natural Disaster Mitigation*, 19–22 February 2004, New Delhi, India, Vol. 2, 447–454.

Waveney District Council. (1993). *Planning for the coast.* Consultation Document. Lowestoft: Waveney District Council.

www.defra.gov.uk/environ/fcd/policy/mrcomp (last accessed 9th January 2006).

www.happisburgh.org.uk (last accessed 9th January 2006).

Chapter 8

The Indian Ocean Tsunami: Local Resilience in Phuket

John Handmer, Bronwyn Coate and Wei Choong

Coastal Vulnerability and Livelihood Recovery

The vulnerability of coastal populations is closely linked to the resilience of their livelihoods. In places where there is little or no public welfare, livelihood resilience or security means the ability to continue to receive income or the resources needed to continue living (Anderson & Woodrow, 1998). Occupants of major coastal cities have most of the facilities, resources and opportunities of city dwellers everywhere. Their livelihoods are dependent on the vitality of the city's economy. But those residing far from major cities may be more directly dependent on their coastal location for a living. The coast may provide for their subsistence directly through fish and similar living resources, or through outsiders who visit to take advantage of the location for leisure, sports, aesthetics, the natural environment or simply entertainment.

The resilience of local livelihoods in many tropical coastal areas depends on tourism. Should the area suffer some major shock, then the longer term effect will be related to the ability of tourism to recover from this impact. The Indian Ocean tsunami of 26 December 2004 devastated many tourism areas including some in Southern Thailand. Resorts and natural attractions were destroyed, many local people and international tourists were killed and the region suffered something approaching the worst possible publicity as countless people searched for their missing friends and relatives against a backdrop of devastation.

Many people lost their lives, but our concern here is with the survivors and specifically how their livelihoods were, and continue to be, affected by the tsunami. Some governments issued travel warnings, urging their citizens to leave the area immediately after the tsunami and to return home, thereby depriving the area and country of desperately needed foreign exchange and employment. But the coastal resorts of Southern Thailand have been working to restore their facilities and welcome tourists back as soon as possible.

One approach to assessing the vulnerability of coastal regions is to examine the impact of a major event on the local economy and on local livelihoods, and their capacity to

Managing Coastal Vulnerability
Copyright © 2007 by Elsevier Ltd.
All rights of reproduction in any form reserved.
ISBN: 0-08-044703-1

Table 8.1: Examples of informal sector occupations in Southern Thailand.

Occupations directly linked to the tourism industry	Occupations indirectly linked to (and supporting) the tourism industry		
Informal tourism/hospitality industry occupations	Informal construction industry occupations	Informal agricultural industry occupations	Informal fishing industry occupations
Touts	Carpenters	Farmers	Fishermen and women
Mobile fruit sellers	Builders	Labourers	Fishmongers
Mobile hawkers	Construction workers	Wholesale food traders	Boat builders
Mobile and retail food vendors	Timber workers	Crop pickers	Net repairs
Craft workers and local artists (tattooists, fire twirlers, artists and wood carvers)	Painters	Small market gardeners	
Market stall holders	Concreters and pavers		
Intermediaries — middlemen (translators)			
Taxi drivers (tuk tuk, taxi and moped)			
Masseuses			
Beach-based beauty therapists			
Sex workers			
Merchandise sellers			
Small local restaurateurs and bars			
Recyclers (plastic and glass collectors)			

recover through bouncing back by adapting to new circumstances if necessary. We have selected the southern Thai island of Phuket as a case study (Table 8.1; Figure 8.1). Although much of Phuket escaped serious damage, the west coast of the island, which is also well known for its tourist luring beauty, was badly damaged by the tsunami.

While much has been reported on the human tragedy, and the pledging of funds of an unprecedented scale, the long-term reality for the survivors of local communities is the struggle to rebuild their lives and livelihoods. This chapter will focus on community resilience and economic recovery in managing the coastal vulnerability that tsunamis present for a region like Southern Thailand. Particular attention is given to the recovery of those engaged in the informal sector of the economy, especially those who have been involved in the tourism industry and other related industries that complement and are

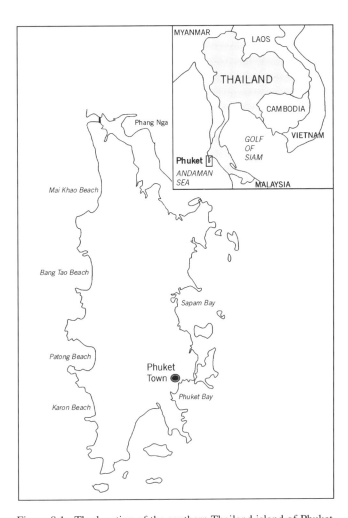

Figure 8.1: The location of the southern Thailand island of Phuket.

interconnected with the tourism industry including fishing, agriculture and construction. A note on fieldwork is at the end of the chapter. This chapter does not deal with the politics of post-disaster international or domestic aid.

The Indian Ocean Tsunami

The 2004 'Boxing Day' tsunami swept more than 8000 km across the Indian Ocean inundating coastal areas, resulting in over 300,000 deaths and enormous physical damage in some areas. The tsunami was generated by a very powerful undersea earthquake that occurred on the morning of 26 December, 2004 off the Northwest coast of the Indonesian island of Sumatra. Estimates of the strength of the earthquake ranged from 9.0 to 9.3, with a wave of up to 10 m.

A tsunami is a natural phenomenon consisting of a series of waves generated when a large body of water in an ocean or lake is displaced. Earthquakes occurring along the ocean floor, undersea landslides and volcanic eruptions, and meteorite impacts have tsunami-generating potential. Tsunamis are frequent global phenomena that generally go unnoticed by coastal people unless their scale and location result in human impacts. Some areas, such as parts of eastern Japan experience them frequently and as a result have substantial engineered protection for coastal towns and developed elaborate warning systems. Traditional knowledge of coastal ocean behaviour is credited with saving the lives of many coastal Mokan people in Thailand and the Andaman Islands.

Coastal Livelihoods through Tourism in Thailand

The land/sea interface is a major focus for global tourism (Miller & Auyong, 1991), and tourist money is the primary source of livelihoods in coastal Southern Thailand. It is not about pleasure, cultural exchange or learning; it is about survival for many of those working in the tourism sector in Thailand. Tourism is largely stimulated by private sector investments, whereby some benefits and opportunities spin off into the informal sector of the economy (see Table 8.1).

There is no disputing the importance of the tsunami-affected region for the Thai tourism industry. According to the Tourism Authority of Thailand (TAT) the six main affected provinces of Phang Nga, Krabi, Phuket, Ranong, Trang and Satun recorded tourism revenue of Baht76 billion from foreign visitors in 2003 (which represents around 25% of Thailand's recorded tourism revenue for the period). In 2004, the tourism industry was conservatively estimated to be valued at US$582.5 million, representing around 12.2% of GDP, employing 8.9% of the workforce (WTTC, 2005). However, these figures reflect the official economy and substantially underestimate the significance of tourism, due to typically encountered problems associated with the capture and measurement of informal sector economic activity in official statistics. The average amount spent per international tourist visiting Thailand in 2003 was US$726, which is high compared with other Asia-Pacific destinations such as Malaysia where the average amount spent by foreign tourists is US$510 (WTO, 2005) (Figure 8.2).

Another important factor when estimating the impact of tourism in local communities and economies, is the leakage that occurs when tourist dollars flow outside the local economy to pay for imports purchased by tourists (such as foreign-owned tour operators, hotels, etc) and also when the profits are directed to those outside the local economy (UNEP, 2002). In Thailand, this leakage from the national economy may be as high as 70% which significantly reduces the benefits from the tourism industry. A leakage rate of 70% indicates that the average amount injected into the national economy per international tourist could be as low as US$218 per visitor (UNEP, 2002), with a substantially lower amount remaining in the Phuket local economy (see Table 8.2). This raises questions for the recovery of the post-tsunami tourism industry and how it can be planned to best contribute to the livelihoods of those in the affected areas. As a general observation, most income generated by the informal sector will circulate in the local economy, while most of that generated by the formal tourism economy — where there is large-scale foreign

Figure 8.2: The Fate of Dollars Spend by Overseas Visitors to Phuket.

Table 8.2: Size of the informal economy in Thailand in 2002.

Estimation of the Size of the Informal Economy	**2.38 Billion Baht**
Share of the informal sector to GDP	43.8%
Share of the informal sector to the labour force	71.9%

investment and control — will leave the local and national economies. In Thailand's case, this leakage should be addressed in recovery planning by acknowledging the contribution of the informal sector to the local economy.

Reducing Disaster Impact by Ensuring Local Livelihoods

In the absence of a functioning economy, those in a disaster-affected area and those dependent on activities in the affected area will depend on external support or welfare for their livelihood and survival. Welfare can come from government, non-governmental organisations (NGOs), international aid and/or family. If there is no welfare they will have to draw on savings or other assets (e.g. see Winchester, 1992), or migrate typically to a major city. We argue therefore that a first priority for recovery efforts should be to support the local economy and ensure that it continues to function by providing for livelihoods and other services in the affected area (see also Handmer & Hillman, 2004). In this context, the economy is not simply that reported in official statistics. It includes the unofficial or informal activities which are often more important for local livelihoods, as explained below. Shanin (2002) argues that three quarters of the world's population relies on the informal economy for their survival. We attempt to examine this important sector, as well as the official and universally recognised economy, as the two are generally closely linked.

Supporting local economic activity may not be straightforward when much of the obvious economy is in the hands of major corporations, and when the post-disaster emphasis is on asset restoration (IFRCRCS, 2001). Although the provision of such support is based more on recovery in poorer economies, the approach can be applied in richer countries, especially in rural communities where aid funds are less likely to recirculate (Handmer & Hillman, 2004).

Disasters may also present opportunities for major change and economic enhancement. Skidmore and Toya (2002) argue that disasters stimulate long-term economic growth, although this appears to be the case primarily for rare earthquake events (Benson & Clay, 2004). There is evidence from the US that even though a local economy may boom following disaster, some sections of the affected community will be substantially worse off (Albala Bertrand, 1993). Similar patterns have been found in poorer economies (IFRCRCS, 2001).

Many analysts argue that ideally, post-disaster restructuring should be about making the local economy (and community) more sustainable (Monday, 2002) — we believe that this is problematic for those whose priority is survival. Within a region such as Phuket, which is dependent on the tourism industry, current asset reconstruction is driven by the long-term anticipation that the tourism sector will recover and new infrastructure will support the recovery of the industry. However, the reconstruction phase is short term and apart from providing income to locals through employment during the tourism downturn the assets themselves that are created and rebuilt will primarily benefit those engaged in the formal sector of the economy.

The Informal Sector in Thailand and the Southern Region

Since Keith Harts' 1971 study of the Ghana labour force (Hart, 1973), much has been written about the informal sector and its often ignored significance for the broader economy. However, there is lack of consensus around its definition and measurement (Allal, 1999; NESDB and NSO, 2004). For simplicity we have defined the informal sector in terms of the nature of the business enterprise, where it is assumed to comprise economic activities that take place outside the framework of corporate, public and registered private sector establishments. Typically the informal sector consists of privately controlled small or micro-scale economic activity that is unregistered with regulatory authorities. Such enterprises usually do not comply with regulations governing labour practices, taxes and licensing requirements. As a result informal economic activity is often seen as representing lost revenue by government tax collectors — who work hard to formalise it. In addition to the tax issue, informal activity may be seen to be competing 'unfairly' with formal activity, selling fakes, and posing a health risk (Cross, 1995).

Informal sector traders often argue that they do not earn enough to enable them to participate in the formal sector (e.g. see Edgcomb & Thetford, 2004). However, there are instances suggesting that some informal sector traders such as Thai moto-taxi, tuk tuk and taxi drivers can manipulate market prices by monopolising services, in what the government has labelled, 'mafia-style' price caps (Phuket Gazette, 2005b). Having the power to set prices can leverage incomes significantly, particularly during high season. Currently, although there is very little work for the transport sector prices remain high, and tourists have little choice but to pay.

A common misperception about the informal economy is that it represents the criminal economy (ILO, 1993). Some activities may be technically illegal but generally tolerated such as the sex industry. We include this sector in our analysis. We do not include households employing paid domestic workers from informal sector enterprises, and treat them separately as part of a category named households according to the definition followed by the ILO (2003).

For much of the world and some sectors within rich countries, understanding the informal economy is the key to understanding people's livelihoods (Shanin, 2002). It is often celebrated by sociologists as showing people's resilience in the face of economic systems that do not offer anything to them. Others, such as the World Bank see the informal sector as something to be eliminated (World Bank, 2005), arguing that it is primarily a tax dodge and connected with over-regulation.

According to the ILO (1993) the informal sector has the following features or characteristics:

- small size of operations
- reliance on family labour and local resources
- low capital endowments
- labour-intensive technology
- limited barriers to entry
- high degree of competition
- unskilled work force and acquisition of skills outside the formal education system.

This list is generally applicable to the informal sector in coastal Southern Thailand. Although skills in the workforce may manifest by informal means in the passing on of cultural traditions and knowledge and through language skills, workers in the formal sector have a higher level of formal educational attainment than those in the informal sector (Allal, 1999). Typically, most workers in the informal sector are involved in small business enterprises employing no more than five persons, in contrast to the formal sector where most are employed in larger scale business organisations. Formal sector employees are generally paid more than their informal counterparts (who however pay no tax), and enjoy a higher degree of job security than those employed in the informal sector (Allal, 1999). They enjoy supplementary benefits such as overtime payments and bonuses, whereas for those in the informal sector supplementary benefits (where these exist at all) tend to be in the form of food supplies and housing. Migrant workers are the worst off in both sectors (MACAW, 2005).

To highlight some of the key areas and occupations within the informal sector that directly and indirectly support tourism Table 8.3 provides an indicative overview. Some of the occupations listed can occur both within the formal and informal sectors, so for instance a subsistence fisherman from a coastal village — who may occasionally take tourists out fishing for cash — would be classified as informal, if he was employed by a resort to do the same things he would be classified as part of the formal sector and would appear in Thailand's GDP.

In Thailand, the informal sector is very important in terms of its contribution to GDP, but what really stands out is the sector's significance in terms of employment as set out in Table 8.3 with just under three quarters of the Thai workforce depending on the informal sector (NESDB and NSO, 2004). By way of comparison, estimates for the US range from 5 to 27% varying hugely by occupation (Losby, Kingslow, & Else, 2003). In Los Angeles County it is estimated that some 29% of the population work informally (Losby et al., 2003).

Table 8.3: Size of the informal sector across selected nations.

Country	Informal Sector (%) of GDP, 1999/2000
Japan	11.3
United Kingdom	12.6
China	13.1
Singapore	13.1
Australia	15.3
Germany	16.3
Canada	16.4
Italy	27.0
Korea Republic	27.5
Greece	28.6
Brazil	39.8
Philippines	43.4
Thailand	52.6[a]

[a] Friedrich Schneider uses the currency demand method to estimate the informal sectors share of national GDP, which in the case of Thailand values the informal sector at 52.6% of national GDP. This estimate is almost 9% greater than the estimate produced by the NESBD and NSO. This illustrates some of the complexity involved in measuring the informal sector and highlights the importance of ensuring consistency in the method of measurement when conducting comparison across nations, which Schneider has done.
Source: Schneider (2002).

To interpret these figures in context, Table 8.3 summarises the relative importance of the informal sector in a range of other countries. In countries without effective government most of the economy will be informal – such as Afghanistan where some 90% of the economy is informal.

While the informal sector is particularly significant in rural areas dominated by agricultural industry, it also accounts for a sizable proportion of economic activity in other sectors within the Thai economy including the tourism industry. In the six southern tsunami-affected provinces of Phang Nga, Krabi, Phuket, Ranong, Trang and Satun that are dependent on the tourism industry, the informal sector is found to account for a conservatively estimated 56% of employment, with the province of Phuket at 31% (NSO, 2005). A likely reason for this may be that the success of Phuket as an international coastal tourist destination has attracted major investment from overseas and from Bangkok — driving small informal sector operators out of the market and formalising the workforce in some areas (although not necessarily always with better pay and conditions).

Evidence from locally based NGOs including the Collaborative Network for the (posttsunami) Rehabilitation of Andaman Communities and Natural Resources, the Coalition Network for Andaman Coastal Community Support and Hi Phi Phi, tend to support the view that the formal and informal sectors do not always sit comfortably together and in particular it appears that tensions often flare over the most limited of coastal resources: the land at the water's edge. Recently the head of the Thai community based at Phi Phi Island, Manop Kongkowreip, expressed the concerns of local people that they would never be able to return to the island if the rumoured plans to turn the island into a luxury resort were true:

> We all want to come back. It's the place where we live, and the place where we were born . . . It's important for the community to stay together (Kongkowreip quoted in McGeown, 2005).

Government authorities have yet to formally announce what the recovery plans are for Phi Phi, but have signalled that Phi Phi is being considered as a 'special area' without specifying exactly what this entails (McGeown, 2005). The combination of rising property prices and the increasing number of new developers and construction activity along the coastal areas (eg: in Kata and Karon beach) is fuelling local community concerns regarding land ownership and affordability of the area — and may thereby undermine local livelihoods. It is also increasing the value of assets at risk from coastal hazards.

The Impact of the Tsunami on the Tourism Industry

Even though human casualties were fewer than those of other Asian countries hit by the tsunami, the cost in terms of human life and injury sustained by both Thai nationals and foreign tourists is significant (Table 8.4).

Aside from the high human cost of the disaster it has been predicted by some economists that Thailand will face high financial costs compared to other countries (McNaughton, 2005); although Raghbendra Jha (2005) of the ANU suggests that the impact will be only about 0.1% of GDP. The highest cost is because of the destruction of infrastructure and property and also the resulting loss of revenues and costs associated with clean up operations. The bulk of wealth generating assets are located in the formal and also private sector and controlled by people outside the local economy. A report in the Phuket Gazette (2005a) highlighted the neglect of public areas over the reestablishment of private sector enterprises at Kamala Beach, due to foreign donations and investment interests.

The financial cost of the tsunami is also relatively high in Thailand because of the country's lucrative tourism industry,[1] which is also a major contributor to the nation's export earnings as foreign expenditures on tourism are counted as export income. The majority of tourists to the region are long-haul international travellers, which has ramifications for recovery given the sensitivity and highly competitive nature of the international tourism market. Media reports and foreign perceptions about a region have the potential to cause further devastation to a disaster-affected destination because of the discretionary nature of travel in which as Cassedy states "the quest for paradise (can) suddenly transform into a dangerous journey that most travellers would rather avoid" (Cassedy, 1991, p. 4). An area

---

[1] The contribution of tourism receipts to GDP varies from around 6–15% according to different sources and also in relation to what areas are included in making up the receipts. The Office of the National Economic and Social Development Board (NESDB) estimate of 6% is based on reported hotel and restaurant, travel agencies and air transport receipts. It is also worth noting that these official estimates will understate the contributions due to the failure of the national accounts to adequately record the impact from the informal sector of the economy.

Table 8.4: The human cost of the tsunami by province in Southern Thailand.

Province	Reported missing	Dead				Injured		
		Thai	Foreign	Unknown	Total	Thai	Foreign	Total
Phang Nga	1758	1242	1633	1349	4224	4344	1253	5597
Krabi	585	357	203	161	721	808	568	1376
Phuket	638	151	111	17	279	591	520	1111
Ranong/Trang/Satun	10	165	6	0	171	322	51	373
Total	2991	1915	1953	1527	5395	6065	2392	8457

Source: WHO (2005). Originally from Department of Disaster Prevention and Mitigation (DDPM), Ministry of Interior, 21 February.

can become stigmatised by a major disaster. For example, many people in the affected areas, particularly in the Kamala Beach area remained convinced of the presence of spirits of the missing, both Thai and foreigner, that haunt the living. Reports began to emerge 10 days after the tsunami as the gravity of the disaster began to sink in. Images of the walking dead and screams in the night have had such a significant impact on the local community that some people are afraid to work in particular areas (Chong, 2005).

This highlights that 'impact' is far from momentary and may be felt for years. Risk transfer mechanisms such as insurance have favoured large-scale commercial enterprises, although insurance payouts have been limited given the extent of devastation (McNaughton, 2005). Many small businesses affected by the tsunami were either not insured, or did not have appropriate coverage (Economist Intelligence Unit, 2005). The informal sector rarely has access to formal financial recovery mechanisms such as insurance and national compensation packages. At the time of fieldwork, it appeared that few, if any, in the informal and small formal enterprise sectors had received any assistance since the tsunami. Many were aware that they could not meet the criteria for claiming aid under the Phuket Action Plan (MACAW, 2005).

The major strategic recovery plan for the tourism sector in Southern Thailand is the Phuket Action Plan. This plan was developed by the World Tourism Organisation with input from leading regional tourism bodies including the Pacific Asia Travel Association (PATA) and the Tourism Authority of Thailand (TAT). The Phuket Action Plan has received widespread support and endorsement at an international and regional level (see Recovery and adaptation strategies).

A central part of the Phuket Action Plan is the positive public relations message that recovery is proceeding well, and that tourists returning to the region are further facilitating the recovery process. Despite the marketing message it appears across both the informal and formal sectors that the tourism industry is still in a slump. The number of international arrivals to Phuket is 67.2% down (PATA/VISA, 2005), well below arrivals for the same period in 2004 (see Figure 8.3). Total number of domestic arrivals is also

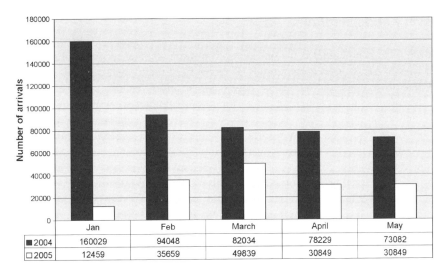

Figure 8.3: Number of domestic and international arrivals at Phuket airport 2004–2005.

down by 42.6% for the same period last year (PATA/VISA, 2005). The average occupancy of hotels is between 10% and 15% which is down from the normal 60% during this annual expected slump (The Nation, 2005). According to data for the first 4 weeks ending 17 July 2005, international inbound spending is also down by 29% (VISA, 2005). Estimates put the loss to Phuket's tourism industry at around 60 billion baht in revenue over the past 6 months (The Nation, 2005). Cooperation and collaborative strategies devised by the TAT, Association of Thai Travel Agents and the Thai Hotels Association appeared to have failed in their attempt to revive the slumping tourism sector in the region. Despite the barrage of specially targeted, cheap package deals advertised, both the actual number of visitors to date since the tsunami along with the expected number of visitors is significantly less than was predicted in earlier post-tsunami assessments, compounding the recovery of other sectors in Phuket (UNRC, 2005).

Evidence from other disasters, which have impacted on the long-haul traveller market, suggests that recovery is often swift once the impacted area has had the chance to clear up and visitors feel assured it is safe for them to return. This was the case in Bali for instance — although the pattern, origin and wealth of the tourists has changed (Regg Cohn, 2005). Tourists have been keeping away from Phuket as they perceive that the clean up is not yet complete and also that it is not yet safe. Safety has been something that Thai authorities have been at pains to emphasis in encouraging foreigners to return to the beaches of Southern Thailand. In doing this they have had to counter the overly cautious travel advisories issued by countries that exaggerate health and safety risks beyond what the World Health Organisation believes to be present and show recovery of the local economies that depend upon the tourism sector (Morison, 2005; WTO, 2005). Further contributing to the slump has been the recent false tsunami warnings alarm which scared many people off the island, and the escalating insurgency in the south of Thailand, have compounded fears by tourists over the safety of the region.

Recovery and Adaptation Strategies

However, what is emerging is the question of how and whether recovery plans will benefit the livelihoods of local communities. If recovery of the tourism sector is structured in a way that always puts the interest of large-scale business ventures ahead of those of the local communities then the benefits of a vibrant tourism industry will be limited for local people and the economies they depend on. We already know that many 'packaged' holidays and more exclusive luxury resorts, where tourists may scarcely venture outside the resort, do little to provide revenue flows back into the local communities — in the absence of strong redistribution policies (UNEP, 2002; see above). This view was reinforced on the ground by many local people complaining that of the small number of tourists circulating the island, few venture out of their resorts and are discouraged from engaging with the local services such as beach masseurs, manicurists and hair braiders. Many hotels will have their own taxi service to the airport and tours around the Island that have cut out the 'middle men' touts who would bring the tourists to their accommodation and transport and introduce them to the sights around Phuket. This has also impacted on informally run restaurants who fail to advertise with popular hotels are doing much less business.

The Formal Sector

Formal sector recovery is guided by the Phuket Action Plan. The Plan focuses on restoring the tourist flows, which generate income for the sector and those whom it employs, rather than simply rebuilding assets (Handmer & Hillman, 2004). It concentrates on saving local tourism jobs, re-launching small tourism-related businesses and recovering the visitor flows that underpin the local economy. Associated marketing aims to restore confidence in the coastal region as a tourist destination (also see Faulkner, 2001). The hosting of the Miss Universe contest in Phuket in May 2005 was seen as a high-profile opportunity to show how the area has rebounded.

The Plan is not about rebuilding infrastructure or hotels, which is largely covered by other agencies, insurance companies and hotel chains. The Phuket Action Plan is built on the principles of sustainable tourism and aims to ensure that the tourism sector emerges from the disaster stronger and more resilient than before, with more environmentally friendly systems, more civil society involvement in the tourism industry and more revenues from tourism remaining in the local community — although it is not clear how these goals will be achieved. A secondary aim is to reduce disaster vulnerability by working with the United Nations Development Program on a tsunami warning system.

A majority of the businesses affected by the tsunami and its impact on the tourism industry are micro and small businesses that are family run. Many are struggling with the loss of family members as well as the collapse in tourism. However, micro enterprises and the informal sector are not specifically mentioned in the Plan, even though these groups appear to have little access to recovery funds. Instead, the community relief aspect of the Plan is centred on assistance to small and medium-sized enterprises (SMEs) such as restaurants, etc. In this context, the Plan is strong on rhetoric but much weaker on useful strategies and activities to assist the SMEs, particularly those in the informal sector. Some local residents, who are now unemployed, have been unable to satisfy the bureaucratic requirements involving the provision of documentation to be able to be classified as 'tsunami affected' and so have effectively been denied access to much assistance. The ambiguity around the definition of 'affected' has been an issue with the recovery Plan. Many residents believe they have missed out because of where they lived in relation to the tsunami's path with the focus being on immediate impact rather than long-term capacity to recover. There is concern that if the tsunami highlighted the acute vulnerability that accompanies financial dependence on the tourism industry, then the tsunami reconstruction plans may exacerbate this even further.

The fifth pillar of the Phuket Action Plan is about risk management, aiming to make coastal tourism destinations safer and more secure from natural disaster with a focus on beachfront construction. However, this emphasis is questioned by many locals as it effectively prevents them from re-establishing their homes where they were before. There is increasing emphasis on land zonation aimed to reduce risk through imposing regulations and codes on where to build and how to build. The land tenure issue which has long plagued local communities in other sectors such as mining, logging and fisheries is now becoming a contentious issue for tourist development and local people. As locations claim value for their resource attractions such as access to the sea, coral formations and scenic beauty, their market value will increase. Since the tsunami, issues of land tenure and the lack of formal documentation of ownership, coupled with increasing demand for prime real estate for economic development, has projected land prices beyond the local affordable price. This scenario is fuelling animosity between private developers and the local community. Instead of restoring 'normalcy', those in both formal and informal tourism sectors will have to learn to adapt to a new set of circumstances.

Informal Sector

Although the Phuket Action Plan emphasises, on paper, local-level engagement via the tourism sector, it appears that there has been little official involvement of local communities, in particular, of the informal sector. There is also the question of addressing the issue of 'leakage' of tourism revenue out of Phuket. While the plan aims to revitalise the tourist industry, the lack of recognition of out movement of desperately needed tourist dollars may significantly slow the local recovery process. Some aid groups have been concerned that lower income groups in both formal and informal sectors, of local or immigrant backgrounds, are unable to receive compensation and assistance; "they should have their voices heard in any related rehabilitation programme" (Friedrich Ebert Foundation, 2005). There is an assumption by those in the informal sector that the recovery of the tourism industry will directly influence their livelihood security. In the meantime, recovery strategies are diverse as people seek new economic activities, move to new areas in search of employment and expand their income sources to survive.

Many people move between formal and informal employment. Some asset reconstruction in Phuket is contributing to the livelihoods of those economically displaced in the informal sector. For instance, it has been observed that the building and construction industry has provided a source of employment to many previously employed in the tourism industry. So, while building materials are predominately imported into the region from mainland Thailand, the labour used in reconstruction has been provided from local informal workers. Some, previously working as touts and tuk tuk drivers are now employed informally without contracts as labourers, painters, builders and concreters earning approximately 100–200 baht a day. Although this is less income than from their previous occupations, and within a lower than average threshold, it is still a significant proportion of household income.[2] Also as property is being acquired by foreign investors to build large up-market resorts which in themselves will have further long-term ramifications for the informal sector and the local community, the building and construction industry is being further propelled albeit temporarily. For those now working in building and construction there is hope that this will carry on long enough to tide them over until the tourism industry recovers and they can return to their usual occupations.

When asked about their recovery in August 2005, interview participants were surprised that anyone would be interested about their lives now that the international media has gone. Although there was significant attention given to affected areas in Thailand, now that the immediate humanitarian needs have been met, the participants felt that no attention was being given to long-term issues. Those involved in the informal sector were aware that they would not be part of the government scheme, but they were becoming increasingly concerned that tourists had not returned to the area. Of the participants, from both the formal and informal sector, none had received any government assistance. There was little mention of NGOs or local organisations, except to point out the new schools being built and the numerous fishing boats that had been donated. None of the participants had been in contact with NGO groups. Recent newspaper reports have documented their situation as the general public in Thailand question what happened to the government-promised funds. Participants mentioned that angry Phuket residents have been protesting in Bangkok over the past few months. No participants will rely upon the government to assist in their direct recovery, and many feel that it is the government's responsibility to boost tourism again. Yet they know that survival and the immediate livelihood security for their families is their responsibility.

In light of the lack of recovery assistance, the linkages and networks of strong kinship ties becomes a source of support both financially and emotionally. Findings reveal that many people in the informal economy with family residing in other provinces in Thailand have greater opportunities to seek employment outside Phuket either on family farms and business, or were able to rely on relatives sending money. However, those with a family base within Phuket were less likely to receive much support as more family members were affected. Participants spoke of sharing food and other resources during the first few months. Now they try to spread the work around between each other, so touts will bring tourists to their taxi driver friends and masseurs. Yet despite the resilience of the informal sector community, they remain worried about the future.

It is clear from the discussion above that the informal sector is largely ignored in practice by government agencies, in spite of being acknowledged in plans, documents and the occasional high-profile political visit. This leaves a third of the workforce to cope through their personal networks and in some cases informal assistance from networks overseas. This is particularly pertinent in light of some of the recent reports coming from some of

---

[2] Based on GDP per capita figures for 2002 (US$7010) and HDI ranking (76) in comparison to other countries in the region. Worst performer, Timor-Leste — 158; best performer. Hong Kong — 23. (UNDP, 2005).

the devastated communities where local residents including minority groups and others referred to as sea gypsies (Molkon) have been locked out of their villages and are under pressure to abandon their settlements along the coast and waterways (CNRACNR and CNACCS, 2005). The area of Khao Lak, north of Phuket is currently experiencing land tenure problems as foreign developers begin to buy up land where the sea gypsies reside on traditional tenure — systematically transformed public land such as beaches into private goods that fence out the local community. As the Molkon construct makeshift shelters by day, they are removed at night by developers. Due to lack of formal documentation, it is becoming increasingly difficult for the local people to recover on the same land they occupied prior to the disaster.

The Phuket Action Plan could be extended to do more for local residents in helping to provide a framework to build strong and resilient local coastal communities. In Gambia a project by the Pro Poor Tourism agency (Bah & Goodwin, 2003) introduced changes that have improved the livelihoods of the local residents from the tourism industry, such as badging of beach vendors, greater marketing and advertising of locally produced handicrafts in hotels, the creation of codes of conduct for informal workers and the establishment of an organisation to represent the needs of the informal sector in negotiating a more cooperative working relationship with the formal sector. Something similar could be introduced to Southern Thailand. By increasing the access of the informal sector to market opportunities in tourism industry, stronger and more resilient local communities will result.

Conclusion — Recovery and Adaptation as Resilience

Coastal areas worldwide are susceptible to tsunami hazard. The recent Indian Ocean experience shows how devastating they can be with high death tolls, physical destruction and the potential to destroy coastal economies dependent on tourism. Following a major impact a key element of managing vulnerability is the adaptation and recovery of local economic activities underpinning local livelihoods. Recovery is not restoration — restoration is not possible after such wholesale destruction, hence the emphasis is on resilience; on flexibility and adapting to new circumstances and new livelihood opportunities.

In assessing local resilience, both the formal and documented — and well publicised — economy as well as the informal economy have been examined. A substantial minority of people in Phuket depend on the informal sector for their livelihoods through generally small-scale enterprises or casual employment. For Thailand as a whole to some extent it supports nearly three times as many people as the formal sector. The informal less visible economy may be more important to local people than the obvious official economic activity and should not be overlooked if we are really interested in reducing vulnerability.

The Phuket Action Plan provides a good framework for the recovery process, and its emphasis on the money flows that employ people and sustain local enterprises is particularly welcome — rather than on asset construction which may give the appearance of a building boom and local economic resilience even though benefits for local people may be limited to short-term construction employment. However, the Plan lacks the specific strategies and 'how to' processes for action that is needed to guide both private and public sector efforts. Discounted packaged holidays to Phuket have had limited success, and unfortunately these packages also support the leakage of profits from the Phuket economy into the hands of largely out of area and overseas interests — and thereby provide very little for informal sector employment — when what is needed most is a revitalisation of the local economy. A priority should be to increase the percentage of the tourist dollar retained locally. Instead it appears that the opposite may be happening. This highlights the gap between rhetoric and the reality of needing secure livelihoods to underpin the rebuilding of people's lives.

Another example of the marginalisation of the informal is provided by the land issue — the land by the sea has very high value for tourist development. In the aftermath of a disaster where land tenure is insecure and undocumented, local people may lose their land. The best they can achieve is a sustainable compromise — most local people do not want massive confrontation as they know that they will inevitably be the losers. When the media and NGO spotlight goes whatever victories they achieve with outside help will be overturned.

What has emerged from Phuket is a three-part recovery process. In the immediate short term, the informal and formal sectors and communities have displayed strength, resourcefulness and resilience in coping with catastrophe. For example, people from the informal sector displayed flexibility and adaptability by shifting sectors to take advantage of short-term employment opportunities in the building and reconstruction boom, by sharing resources and work and by harnessing their kinship networks.

Now, in the medium term, as people are just able to survive from day to day, they remain optimistic about the future as they wait for the local economy to recover. The resilience of coastal communities in Southern Thailand is supported by their ability to rely upon strong social networks consisting of kinship ties and community linkages both within the immediate disaster area, within Thailand and abroad. Although the government and a number of NGOs have attempted to assist some of the informal sector, such as fishing communities, there has been little effort to recognise and encourage the recovery of the local economy and those informal sector workers who contribute to and support it. Despite such a heavy reliance upon a single industry, the tsunami has revealed that these communities have strong social systems and the ability to adapt to new circumstances that enables them to cope in some of the most devastating situations. The people are waiting out the normal quiet season, which has been severely exacerbated by the tsunami, and hoping for a return of the tourist masses during the coming high season that will mark the start of long-term recovery for the formal and informal sectors of Phuket.

Note on Fieldwork

Field work was conducted between 31st July and 3rd August 2005 in Phuket. Participant observations, informal interview(s) and discussions were used to collect qualitative data, with particular emphasis on targeting the informal sector. Participants were mainly from the western coast of Phuket from Bang Tao, Kamala Beach, Surin Beach, Patong, Karon and Kata Beaches. The interviewer was taken along the western coast by a tout and introduced to various people including tuk tuk, taxi and moped taxis, masseurs, restaurant owners, hotel workers, bar staff and fishmongers. The interviewer then travelled south towards Kata Beach and met various beach boys, bar staff, young artists, tuk tuk drivers, beach manicurists and masseurs, sex workers and hospitality staff. Interviews and discussions were conducted in an informal manner with the aim to investigate the current socio-economic status of the participants, how it differed from the previous year (prior to the tsunami), how people have survived over the past 6 months, who they rely upon for employment, financial and emotional support and how they see their future. All participants were willing and happy to discuss their situation. They were mostly open and very friendly. People were surprised that anyone would be interested in their present situation and future plans.

References

Albala Bertrand, J. M. (1993). *The political economy of natural disasters with special reference to developing countries*. Oxford: Clarendon Press.

Allal, M. (1999). Micro and small enterprises (MSEs) in Thailand – definitions and contributions. In: G. Finnegan (Ed.), *Micro and small enterprise development and poverty alleviation in Thailand*. Working Paper 6. Bangkok International Labour Organization.

Anderson, M. B., & Woodrow, P. J. (1998). *Rising from the ashes: Development strategies in times of disaster*. London: Intermediate Technologies Publications.

Bah, A., & Goodwin, H. (2003). *Improving access for the informal sector to tourism in The Gambia*. Pro Poor Tourism (PPT). Working Paper No.15. Economic and Social Research Unit (ESCOR), UK Department for International Development (DFID).

Benson, C., & Clay, E. (2004). *Understanding the economic and financial impacts of natural disasters*. Washington, DC: World Bank.

Cassedy, K. (1991). *Crisis management planning in the travel and tourism industry: A study of three destination cases*. San Francisco: Pacific Asia Travel Association.

Chong, T. (2005). Ghosts stalk Thai tsunami survivors. BBC News, 25 January. Available online http://news.bbc.co.uk/1/hi/world/asia-pacific/4202457.stm (last accessed 2nd March 2006).

CNRACNR and CNACCS. (2005). Policy recommendation on the rehabilitation of Andaman small-scale fisherfolk communities: Tsunami aftermath. Collaborative Network for the Rehabilitation of Andaman Communities and Natural Resources & The Coalition Network for Andaman Coastal Community Support. Available online http://www.icsf.net/jsp/english/flashnews/rehabDocs/tha0401.doc (last accessed 2nd March 2006).

Cross, J. (1995). Formalising the informal economy: The case of street vendors in Mexico City. Available online www.openair.org/cross/vendnow2.html (last accessed 2nd March 2006).

Economist Intelligence Unit. (2005). *Asia's tsunami: The impact*. Dartford: Patersons.

Edgcomb, E., & Thetford, T. (2004). *The informal economy: Making it in rural America*. Washington, DC: The Aspen Institute.

Faulkner, B. (2001). Towards a framework for tourism disaster management. *Tourism Management*, 22, 135–147.

Friedrich Ebert Foundation. (2005). Tsunami impact on Thai workers. Available online http://www.fes-thailand.org/archiv/tsunami/tsunami.htm (last accessed 2nd March 2006).

Handmer, J., & Hillman, M. (2004). Economic and financial recovery from disaster. *Australian Journal of Emergency Management*, 19(4), 44–50.

Hart, K. (1973). Informal income opportunities and urban employment in Ghana. *The Journal of Modern African Studies*, 11(1), 61–89.

IFRCRCS. (2001). *World disaster report*. International Federation of Red Cross and Red Crescent societies, Geneva: International Federation of Red Cross and Red Crescent (IFRCRC).

ILO. (1993). Proceedings of the 15th international conference of Labour Statisticians (ICLS): Highlights of the conference and text of the three resolutions adopted. *Bulletin of Labour Statistics*, 2, IXXXIV.

ILO. (2003). Guidelines concerning a statistical definition of informal employment. *Proceedings of the 17th International Conference of Labour Statisticians (ICLS)*. November–December 2003, Geneva. Available online http://www.ilo.org/public/english/bureau/stat/download/guidelines/defempl.pdf (last accessed 2nd March 2006).

Jha, R. (2005). The Asian tsunami disaster – the prospects for recovery. ANU Reporter, Australian National University, Autumn, 2005.

Losby, J. L., Kingslow, M. E., & Else, J. F. (2003). *The informal economy: Experiences of African Americans*. Washington, DC: The Aspen Institute.

MACAW. (2005). *Tsunami impact on workers in Thailand. Mobile assistance centre for affected workers*. Bangkok: Friedrich Ebert Stiftung Foundation.

McGeown, A. (2005). Rebuilding Thailand's island dream. BBC News, 18th March. Available online http://news.bbc.co.uk/1/hi/world/asia-pacific/4361541.stm (last accessed 2nd March 2006).

McNaughton, T. (2005). The economic impact of the Asian Tsunami. 4th January 2005. BT Financial Group. Available online http://www.btfinancialgroup.com/downloads/commentaries/bto_tsunami_impact.pdf (last accessed 2nd March 2006).

Miller, M. L., & Auyong, J. (1991). Coastal zone tourism. A potent force affecting environment and society. *Marine Policy*, 15(2), 75–99.

Monday, J. (2002). *Building back better: Creating a sustainable community after disaster*. Natural Hazards Informer 3, Boulder: University of Colorado.

Morison, A. (2005). Scare tactics will do far more harm than good. The Age, 30th January. http://www.theage.com.au/news/Opinion/Scare-tactics-will-do-far-more-harm-than-good/2005/01/29/1106850157941.html (last accessed 2nd March 2006).

NESDB and NSO. (2004). The measurement of the non-observed economy in Thailand national accounts. Paper prepared for the United Nations economic and social commission for Asia and the Pacific workshop on assessing and improving statistical quality: Measuring the non-observed economy. 11–14 May, Bangkok. National Economic and Social Development Board and National Statistics Office (Eds.).

NSO. (2005). Labour Force Survey, National Statistical Office, Economics and Social Statistics Bureau, Thailand. Available online http://web.nso.go.th/eng/stat/lfs_e/lfse.htm (last accessed 2nd March 2006).

PATA/VISA. (2005). Tsunami destinations – overall. VISA/Pacific Asia Travel Association Post Tsunami Update, August 2005.

Phuket Gazette. (2005a). Seven months after the wave, why is Kamala still like a garbage tip? July 30–August 5.

Phuket Gazette. (2005b). Governor lays out strategy for war on 'taxi mafia', 10 June. Available online http://www.phuketgazette.net/news/index.asp?fromsearch=yes&Id=4346 (last accessed 2nd March 2006).

Regg Cohn, M. (2005). What Bali can teach Phuket. *Toronto Star*, February, Canada.

Schneider, F. (2002). *Size and measurement of the informal economy in 110 countries around the world*, World Bank, July (available online: http://rru.worldbank.org/documents/paperslinks/informal_economy.pdf).

Shanin, T. (2002). How the other half dies? *New Scientist*, August 3.

Skidmore, M., & Toya, H. (2002). Do natural disasters promote long run growth? *Economic Inquiry*, 40(4), 664–687.

The Nation (2005). Post-tsunami tourism: Phuket struggles to reverse slump, 27 June. Available online http://www.nationmultimedia.com/search/page.arcview.php?clid=6&id=117591&date=2005-06-27&usrsess=\ (last accessed 2nd March 2006).

UNDP. (2005). *International cooperation at a crossroads: Aid, trade and security in an unequal world*. Human Development Report 2005, New York: United Nations Development Program. Available online http://hdr.undp.org/reports/global/2005/ (last accessed 2nd March 2006).
UNEP. (2002). Economic impacts of tourism. United Nations Environment Programme. Available online http://www.uneptie.org/pc/tourism/sust-tourism/economic.htm (last accessed 2nd March 2006).
UNRC. (2005). *Disaster field situation*. Report 16 from United Nations Resident Coordinator (Thailand), 7 July.
Winchester, P. J. (1992). *Power, choice and vulnerability: A case study in disaster mismanagement in south India*. London: James and James.
World Bank. (2005). The informal economy and LED (local economic development). The World Bank Group. Available online http://www.cipe.org/programs/informalsector/articles/wbpapers.htm (last accessed 2nd March 2006).
WTO. (2005). *Tsunami relief for the tourism sector: Phuket Action Plan*. World Trade Organisation, World Tourism Organization Press. Available online http://www.world-tourism.org/tsunami/Phuket/Draft%20Phuket%20Action%20Plan-A%20Rev.3.pdf (last accessed 2nd March 2006).
WTTC. (2005). *Thailand: Travel and tourism report*. World Travel and Tourism Council. Available online http://www.wttc.org/2005tsa/pdf/Thailand.pdf (last accessed 2nd March 2006).

Chapter 9

Vulnerability of the New York City Metropolitan Area to Coastal Hazards, Including Sea-Level Rise: Inferences for Urban Coastal Risk Management and Adaptation Policies

Klaus Jacob, Vivien Gornitz and Cynthia Rosenzweig

Introduction

Many of the world's largest cities are situated at coasts and in estuaries at or near sea level. Major coastal urban centers have long been vulnerable to natural hazards, such as storm surges, shoreline erosion, or even the occasional destructive tsunami (Nicholls, 1995). By the end of this century, increased rates of sea-level rise (SLR) could cause permanent inundation of portions of low-lying coastal cities, repeated flooding episodes, and more severe beach erosion (Houghton et al., 2001; McCarthy, Osvaldo, Canziana, Dokken, & White, 2001). The anticipated SLR will challenge coastal managers and decision makers to adapt to and mitigate these potentially adverse effects of climate warming in innovative and creative ways.

The vulnerability of the New York City metropolitan region to SLR was examined as part of the Metropolitan East Coast (MEC) Report for the National Assessment of Potential Consequences of Climate Variability and Change for the United States (Gornitz, 2001; Jacob, 2001; Rosenzweig & Solecki, 2001; Gornitz, Couch, & Hartig, 2002). The region can be considered as an example of a megacity of global importance in international business, finance, trade, culture, education, and diplomacy. The combined New York City/MEC region's role as a megacity is closely linked to its highly developed infrastructure, particularly an efficient and reliable public transportation system. Economic activity, public safety and health depend on growth and modernization of its complex infrastructure. Appropriate responses or adaptations to changing circumstances, including climate change, are essential in maintaining this region's global position.

The greater New York City Metropolitan East Coast area (MEC) encompasses an area of 33,670 km^2, and 22 million inhabitants of which around 8 million reside in New York City

142   Klaus Jacob et al.

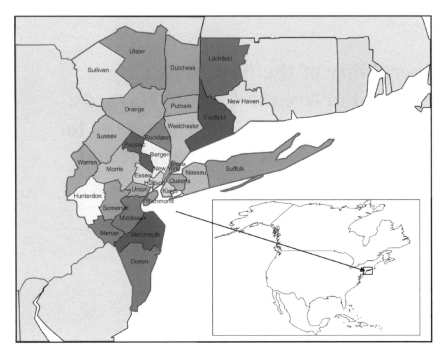

Figure 9.1: Map of Metropolitan East Coast Study Region with insert location. Thirty-one counties are indicated by name; 14 are located in the State of New York (five of which constitute New York City), 14 in New Jersey and three in Connecticut. For details see text.

proper. The definition of the MEC region adopted here is based on work-related commuter patterns moving a large work force to and from the central business district, largely in Manhattan, New York City. The so-defined MEC region (Figure 9.1) consists of 31 counties in three states (New York, NY; New Jersey, NJ and Connecticut, CT). Fourteen counties are located in NY State, five of which constitute New York City, 14 in New Jersey, and three in Connecticut.

With over 2000 km of shoreline, the region's development has historically been closely linked to the sea. Over 2000 bridges and tunnels exist in New York City alone, and many of the larger bridges connect the four (out of five) New York City island boroughs with each other and the mainland. High-density commercial and residential development is rapidly replacing abandoned factories and piers along the waterfront in metropolitan New York and New Jersey, as is happening in many other coastal cities that have moved from a manufacturing to a service industry-based economy. Mid-town and Lower Manhattan are two of the world's major financial centers. Plans are underway for the redevelopment of the (flood prone) World Trade Center site in Lower Manhattan, and of the Brooklyn waterfront, long home to the former Brooklyn Naval Shipyards. New vacation and year-round houses are under construction on barrier island dunes on Long Island, NY, and the northern New Jersey shore. Beaches and coastal wetlands provide recreational opportunities for urban populations and critical habitat for wildlife and fisheries. These littoral environments are caught between the twin pressures of development and increasing coastal hazards.

The south shore of Long Island is flanked by a string of barrier islands and beaches that developed after the end of the last ice age, as glacial sands and gravel were eroded and deposited into ridges and shoals offshore. The present barrier islands are only a few thousand years old, while the ancestral islands lay lower and seaward of their present positions. Most of the southern Long Island shoreline has been eroding over the last 150 years (Leatherman & Allan, 1985), associated with the historic SLR (Leatherman, Zhang, & Douglas, 2000; Zhang, Douglas, & Leatherman, 2004). Major erosion has continued or even accelerated *after* emplacement of jetties to stabilize several inlets, between the 1940s and 1960s, and construction of groynes near Westhampton, on eastern Long Island, in the late 1960s, which curtailed the westward longshore drift of sands. Similarly, the northern New Jersey coastline has tended to retreat since the 1830s, with increased rates of erosion at several localities following erection of "hard" structures. The United State Army Corps of Engineers has spent $2.6 billion (1996-valuation) nationally and $884 million within the Tri-State (New York, New Jersey, and Connecticut) region on beach nourishment costs starting in the late 1920s, with rapid cost accelerations since the 1950s. Over $250 million was spent at just the six sites investigated for the MEC report (Gornitz et al., 2002).

The MEC region is affected by extra-tropical cyclones ("nor'easters") that occur largely between late November and March, and less frequently by tropical cyclones ("hurricanes") that typically strike between late July and October. A map of potential "worst-track" flooding scenarios for hurricanes of Saffir–Simpson (SS) scale 1–4 is shown for lower Manhattan (Figure 9.2).

Figure 9.2: Expected zones of storm Surge flooding in lower Manhattan and parts of Brooklyn as a function of storm level on the Saffir–Simpson Scale (SS 1–4). For details see text.

Large tracts of lower Manhattan would be flooded, dependent on the storm's level. Figure 9.2 shows the increase in flooded areas as a function of storm severity from SS = 1 to 4. Only the dark-shaded areas at the core of central and lower Manhattan are believed to be free currently from modeled flood hazards. Note that SS = 5 storms are currently thought to be unlikely in these mid-latitudes under current climate conditions. Shown flood zones assume sea level as of the year 2000. Flooding becomes more pervasive as SL rises. The "worst-track" storm surge for SS = 4 (lightest pattern) would be associated with a cresting surge height of about 10 m near the *Battery*, at the southern tip of Manhattan, and diminishes in amplitude upstream (northward). Note that for worst-track storms of SS = 3 and higher, lower Manhattan would be split into two islands in the vicinity of Canal Street, and would isolate the lower Manhattan "Financial District" (Wall Street, New York Stock Exchange, etc.). Such storm scenarios would require large lead times (at least 8 h) to achieve safe evacuation of the large office work force, especially since subway and vehicular tunnels are likely to become flooded, and major bridges may have to be closed because of high winds hours before the eye of the storm passes. The destructive potential of hurricanes arises from the combined effects of high winds (>120 km/h), heavy rainfall, and coastal flooding due to storm surge and waves. The flood height is magnified if the surge coincides with high tide. Although wind speeds of nor'easters are lower than those of hurricanes, they are capable of causing significant damage because duration is normally longer. The longer storm surge duration (days vs. hours during generally faster moving hurricanes) allows flooding to penetrate farther inland, and thus may cover a broader areal extent. The nor'easter of December 1992 produced some of the worst flooding in the New York City metropolitan area in 40 years, resulting in an almost complete shutdown of the regional transportation system and evacuation of many seaside communities. This storm revealed the vulnerability of the regional transportation system to weather-related disruptions. Most area rail and tunnel entrance points as well as the three major airports lie at elevations of 3 m or less above the locally still used reference mean sea-level datum of 1929 (U.S. ACOE/FEMA/NWS, 1995; U.S. Jacob, 2001). Flood levels of only 0.3–0.6 m above those produced during this storm could have led to even more severe flooding and to loss of life.

Neither Atlantic basin hurricanes nor extra-tropical cyclones show as yet any proven long-term trends in frequency, strength or spatial patterns in response to climate change. However, they do exhibit considerable inter-decadal variability. In particular, hurricanes may be entering again a more active period (Elsner, Jagger, & Niu, 2000; Zhang, Douglas, & Leatherman, 2000; Goldenberg, Landsea, Mestas Nuñez, & Gray, 2001) during the first two decades in the 21st century. Regardless, flooding due to coastal storms is likely to become more commonplace with rising sea levels, as the surge height will be superimposed on the higher ocean level.

We now review the vulnerability of the New York City metropolitan region to SLR, based on the findings of the MEC report. We outline relevant information and research needs to develop or improve the framework for coherent adaptation and coastal management policies facing the effects of global warming, and in particular rising sea levels.

Vulnerability of the New York City Metropolitan Area to Sea-Level Rise

Impacts of rising sea levels for selected localities within the MEC area were investigated. A suite of five plausible scenarios was used (Figure 9.3), based on extrapolation of historic

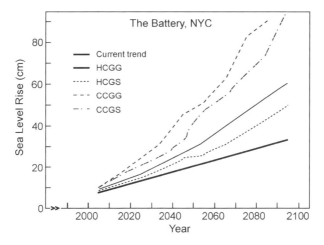

Figure 9.3: Five models of projected sea-level rise for the Battery at the southern tip of Manhattan, New York City. The models and abbreviations are explained in the text.

sea-level trends from tide gauge data and projections from the two global climate models (United Kingdom Hadley Centre, HC; Canadian Centre for Climate Modelling and Analysis, CC). In Figure 9.3, the models generated by the two centers (HC and CC), come each in two versions marked as GG and GS. GG stands for models that only considered the warming effects from green house gases, while GS stands for a modification of the GG models by considering in addition the slight cooling effect of sulfate aerosols in the atmosphere due to reflection and scattering of solar radiation. Note that the four models shown and the extrapolation of the historic SLR trend curve all include the local isostatic subsidence (Peltier, 2001).

The five SLR curves in Figure 9.3 use as a zero-baseline the averaged mean sea level for the period 1961–1990.

The SLR scenarios were coupled with U.S. Army Corps of Engineers surge (WES Implicit Flooding Model) and beach nourishment models (SBEACH; Bruun rule) (Gornitz et al., 2002). Storm surge probabilities were calculated at high tide for combined effects of hurricanes and nor'easters, assuming *no change* in storm frequency due to climate change, and *excluding* wave effects, both of which would worsen the flooding scenarios.

Mean global sea level has been increasing by 1–2 mm/yr for the last 150 years (Houghton et al., 2001), with 1.8 mm/yr a "best estimate" for the last 50 years (Church, White, Coleman, Lambeck, & Mitrovica, 2004). In the MEC region, observed 20th century rates of relative SLR range between 2 and 4 mm/yr, with an average value of 2.7 mm/yr for New York City (NOAA/NOS, 2005) since the 1850s, and a 10% higher mean rate for just the 20th century. Regional sea-level trends are somewhat higher than the global mean because of coastal subsidence in response to glacial isostatic readjustments (Peltier, 2001). By the 2080s, regional sea levels could climb by 0.24–1.08 m (Figure 9.3) above late 1980 levels, i.e. over a period of 100 years. More importantly, flood heights for the 100-year coastal storm could attain 3.2–4.2 m above the 1929 reference datum. Currently, the 100-year flood height above the same reference datum for New York City is 2.96 m — very close to the area shaded in gray in Figure 9.4.

146   Klaus Jacob et al.

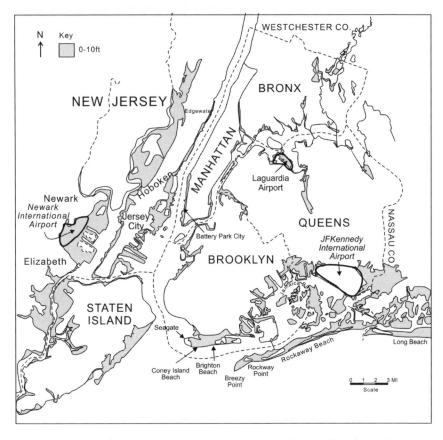

Figure 9.4: Map of the central portion of the MEC study area. Gray shading shows the areas at elevations below 3 m (10 ft to be exact) above the present mean sea level.

The zone within about 3 m above current sea level depicted in Figure 9.4 measures on the order of 300 km². This represents more than 10% of the total shown land area, and encompasses portions of lower Manhattan (New York County), coastal areas of Brooklyn (Kings County), Queens, Staten Island (Richmond County), Nassau County, NY, and the New Jersey Meadowlands (mostly in Bergen County, NJ, see Figure 9.1). Owing to SLR these areas could experience a marked increase in flooding frequency. For instance, the recurrence interval of the 100-year storm flood could shorten to as little as 4–60 years (Figure 9.5). The tidal wetlands of Jamaica Bay between Brooklyn (Kings County) and Queens, which provide prime habitat for migratory birds, already face serious losses and could disappear altogether with rising sea levels (Hartig, Gornitz, Kolker, Mushacke, & Fallon, 2002). Beach erosion rates could increase by 4–10 times over present rates (Gornitz et al., 2002). This would necessitate up to 26% additional sand replacement by volume (and associated costs) on beaches, due to SLR alone. Economic losses from storm flooding and inundation is expected to triple (see below).

Vulnerability of New York City to Coastal Hazards 147

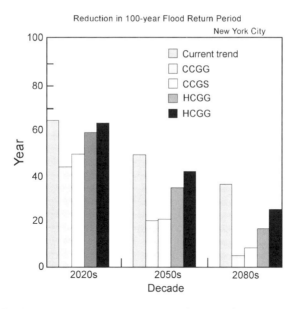

Figure 9.5: Reduction in the 100-year recurrence period for three future decades and the five sea-level rise models as shown in Figure 9.3.

Many elements of the regional transportation system, other essential infrastructure, such as sewage and wastewater treatment plants, as well as commercial and residential property lie at elevations of 2–6 m above present sea level — well within the range of projected surges for tropical and extra-tropical cyclones (Jacob, 2001). Even the seemingly modest increase in sea level of up to 1 m by the end of the century would raise the frequency of coastal storm surges and related flooding by factors of 2–10, with an average of around 3. This would place many public facilities, and especially low-lying critical elevations of many transportation systems, at ever more frequent flood hazards (Figure 9.6).

The rate of losses incurred by the entire region from storms and coastal floods would increase correspondingly with the increased frequency of flood hazards (Table 9.1).

Anticipated annualized average losses due to these floods, on the order of $1 billion/year, appear to be relatively small compared to the annual $1 trillion (2000) regional economy. But in reality major losses do not occur in regular annual increments. Instead, they result largely from less frequent, high-magnitude, extreme events. Extreme storm losses can be expected in this region to exceed $100–200 billion in some cases. During the recovery period, the regional economy could show signs of strain, local businesses could close, and insurers would be stretched to the limit. The approximate $100 billion loss to the regional economy in the aftermath of the September 11, 2001 terrorist attacks on New York City are expected to adversely effect the MEC economy for at least a decade. Nearly half of the 2001 losses appear to be insured losses. The insurance industry was ill prepared and severely strained after losses associated with hurricane Andrew that hit Florida in 1992 and caused total losses of about $20 billion of which less than half were insured losses. The 2004 hurricane season with four major hurricanes originating in the Atlantic that hit

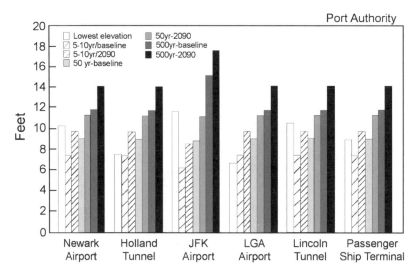

Figure 9.6: Current lowest critical elevations of facilities operated by the Port Authority of New York and New Jersey vs. changing storm elevations at these locations for surge recurrence periods of 10, 50 and 500 years between 2000 (baseline) and the 2090s. See text for details. *Note*: 10 ft equal approximately 3 m.

the U.S. coasts produced *insured* losses currently estimated to exceed $20 billion, despite the fact that none of the hurricanes had landfalls near a major city. Especially New Orleans (a large portion of which lies *below* sea level behind levees) escaped narrowly the path of the 2004 hurricane *Ivan*, but was squarely hit by Katrina in 2005. Risk consultants to the insurance industry advised that the 2004 losses from hurricanes were not unusual and that insurers should be prepared to deal with similar or even higher losses on a regular basis (Anonymous, 2004). The same source states that historically the year 2004 is one of 8 years in the last century that had hurricane-related insurance losses exceeding $20 billion, when adjusted to 2004 valuations. With climate change and SLR and with expanded coastal development, these losses are expected only to rise. The forecasts materialized promptly in 2005 with combined economic losses from Katrina and Rita estimated $140 billion ($65 billion insured).

Coastal Risk Management and Adaptation Issues

Science and Technical Needs

Coastal managers and other public officials need accurate, reliable, realistic information about current and future climate, as well as socioeconomic trends upon which to base their decisions. While the MEC report represents an important step forward in assessing climate change impacts in a major urban area, a number of issues were identified that require further

Table 9.1: Estimates of losses (in 2000-US$) in the MEC region for storms with shown surge heights.

Equivalent Saffir-Simpson category[a]	Surge height[b]		Surge recurrence period (years) in		Estimated total losses (billon $)	Annualized losses (million $/yr)	
	(ft)	(m)	2000	2100		2000	2100
Extratopical storm	8	2.4	20	6	1	50	170
1	10	3.1	50	15	5	100	330
2	11	3.4	100	30	10	100	300
3	13	4.0	500	150	50	100	300
3–4	14	4.3	1000	300	100	100	300
4	16	4.9	2500	800	>250	100	300
All storm categories combined						500	1500

Note: The surge recurrence periods shorten from 2000 to 2100 by an average factor of 3 due to SLR alone, and annualized losses roughly triple. The exposed asset values and storm frequencies are assumed to remain at the 2000 levels. Actual increase in assets and storm frequency would further increase losses. For details see text.
[a]Use only the year 2000 recurrence period for this first column, since the study assumed the frequency of storms would remain the same, and only the surge frequency for same surge height would shorten due to SLR.
[b]Surge height above the National Geodetic Vertical Datum (NGVD) of 1929, which then represented approximately local mean sea level.

investigation, because of limitations in data availability, uncertainties in the models used, or incomplete understanding of basic physical and socioeconomic processes. Reducing uncertainties associated with future SLR requires better knowledge of heat penetration into the oceans (e.g. Levitus et al., 2001), the resulting thermal expansion (Cabanes, Cazanave, & Le Provost, 2001), rates of mountain glacier melting (Dyurgerov & Meier, 2000), and the likely contributions of the Greenland and Antarctic ice sheets (Gregory, Huybrechts, & Raper 2004; Thomas et al., 2004). Furthermore, we need to be able to anticipate changes in tropical and extra-tropical storm frequencies and intensities, and how such changes will affect coastal flooding and beach erosion. New physically based models need to be developed to relate the shoreline's response to SLR. Existing models are often based on empirical relationships, which in turn are based on oversimplifications of incompletely understood complex physical processes (Thieler, Pilkey Jr, Young, Bush, & Chai, 2000). Other important effects of SLR not fully understood in their consequences are upstream migration of the Hudson River salt front and its effects on the Chelsea Pump station, an emergency source of drinking water for New York City during periods of drought. The pump station is located about 90 km upstream from New York City. Related issues are the infiltration of saltwater into already stressed Long Island aquifers, and the effects of salinity changes on the estuarine ecology.

Other types of information needed for improved impact assessment include: more detailed topographic data down to 10-cm resolution over land, and 1-cm resolution near sea level; more detailed data on recent and historic storm damages; accurate inventories of major infrastructure components, their fragility with respect to storm surge, flooding, and wind hazards; their monetary value; and finally, data management and improved storm surge damage modeling capabilities such as that provided by the HAZUS-MH-MR1 tool (FEMA, 2005). The HAZUS multi-hazard (MH) loss assessment tool uses *Geographical Information Systems* (GIS) and currently allows us to quantify damage levels, physical, financial and economic losses, and other impacts from three types of hazards: wind, (riverine and coastal) flooding, and earthquakes. HAZUS can be applied to individual deterministic scenario events, or probabilistically. The latter option allows obtaining annualized losses. HAZUS and similar risk and loss assessment tools have been used for about a decade by the insurance industry for portfolio risk management, and by federal, state, and local governments for emergency planning and disaster-response management. These tools undergo continuous refinements and require time-consuming efforts to keep pertinent data bases updated on hazard assessments, changing asset inventories, and asset fragilities/vulnerabilities to the various hazards.

The stated scientific and technical needs require concerted and sustained investments into financial and personnel resources. The academic/professional research communities, the private business sector, and the various levels of government must share commitments. It requires especially close cooperation and partnerships by many stakeholders that operate large infrastructure/utility systems, whether privately or publicly owned and managed. Their databases and technical know-how are invaluable to assessing and managing the coastal risks and vulnerabilities realistically and effectively. The MEC study experienced mixed results, with both successes and failures to achieve these partnerships and establish a sustained network of active and knowledgeable professionals that could advance the scientific and technical base for assessing and managing the coastal vulnerabilities in the region.

Responses to Coastal Hazards

A number of local, state, federal and private agencies are responsible for responses to natural disasters. The National Weather Service (NWS) of NOAA routinely tracks storms by satellite and furnishes hurricane or storm flood warnings. The NWS works closely with the New York/New Jersey/Connecticut State Emergency Management agencies and the New York City Office of Emergency Management to assess the situation at a local level. These state or city agencies then decide whether or not to declare a storm emergency and whether to recommend closure of government offices, schools, and private businesses. If necessary, evacuation of low-lying areas and beaches via prescribed evacuation routes to emergency shelters is ordered. The Federal Emergency Management Agency (FEMA) provides financial aid for reconstruction efforts. FEMA's National Flood Insurance Program (NFIP) underwrites flood insurance to communities that adopt measures to reduce future flood risks in hazardous areas (FEMA, 1997). U.S. Congress passed Public Law 106–390, the "Disaster Mitigation Act of 2000" to strengthen FEMA assistance to communities for mitigation measures, especially for properties in flood zones where past repeated losses had occurred, while limiting *future* disaster assistance if after *repeated* flood disasters, mitigation measures have not been undertaken. NFIP also specifies designation of erosion zones and setbacks or buffer zones for highly vulnerable coastal areas. The U.S. Army Corps of Engineers builds and manages dams and levees to minimize flood damage. They also undertake beach "nourishment" and some tidal marsh restoration projects. Although the Army Corps factors *historic* SLR rates into their projections of sand volumes needed for beach nourishment, they so far have not considered the possibility of future *accelerated* SLR. Despite these general measures, often on the federal level, development pressures on the local level continue to place even new assets in flood prone areas, or in areas that will become flood prone in the future due to rising sea level.

Organizations and institutions are more likely to react to rapid-onset hazards lasting several hours or days (e.g. flooding from storm surges), rather than to slow-onset hazards that develop over longer time periods (e.g. coastal erosion and SLR). Measures to reduce vulnerability to future coastal hazards, such as SLR, should build upon already-existing programs and institutional mechanisms. For example, SLR projections should be incorporated into the design, siting, and construction of new or updated facilities. An example for an *obstacle* to such forward-thinking, preventive measures is NFIP's *flood insurance rate mapping* (FIRM) program. FIRM flood-zone maps are used for land use planning and construction regulations by many local jurisdictions. But FIRM maps do not yet recognize the *future* contributions in areal (and vertical) extent of flood zones due to any SLR, not to speak of future accelerated rates of SLR. Therefore, FIRM maps in coastal zones, even those currently produced under the current FIRM map modernization program costing in excess of U.S. $1 billion, will become outdated during just a few decades. Many new investments in or near the currently defined coastal flood zones (which ignore future SLR) will suffer ever-increasing losses. Other examples of missed or successful adaptation measures are provided below.

One adaptation option that does not protect the outer shores and barrier islands, but is intended to protect the New York–New Jersey Harbor estuary, has been recently proposed and would consists of three strategically placed storm surge barriers. This concept is in the earliest stages of scientific exploration (Bowman et al., 2004) of its technical and environmental implications. Therefore, its economic, cost/benefit and political feasibility and long-term environmental impacts are as yet entirely unknown. It would be a capital-intensive "structural" solution (meaning an engineered solution) that would provide *physical* defenses against the hazards from SLR and storm surges. To some extent it would therefore implicitly promulgate the otherwise unsustainable waterfront land-use and development policies of the past and present. This approach to adaptation would be in contrast to any "non-structural" solutions that would curtail waterfront development and require changes in land use and zoning regulations, perhaps even relocation or raising of structures and infrastructure systems near the present waterfront. The barrier solution would build on the experiences gained with similar storm surge protection systems built in the Netherlands, and across the Thames River near London. Preliminary hydrological modeling results are given by Bowman et al. (2004). It is too early to tell whether such structural approaches have any merit or chance of future realization, especially since they only protect assets and people in the inner harbor, but not those directly exposed along the Atlantic coast or the Long Island Sound. The public is only gradually beginning to confront implications from SLR and other effects of climate change for the New York City metropolitan region. It is uncertain at this time whether the public process toward any expensive mitigation measures, whether structural or non-structural, can be politically advanced without the region first having to experience a catastrophic storm surge disaster. In the Netherlands and England the tragic 1953 North Sea storm floods triggered there the respective flood protection engineering projects. Currently the U.S. is in a lively discussion (Mileti, 1999) whether non-structural (land use, zoning, and code) measures are better suited than structural (dam, levee, and barrier) measures to provide sustainable protection for communities. The focus of the discussion is on economically sustainable protection, i.e. short-term vs. the long-term measures, especially since there are no well-known upper limits for SLR when projecting many centuries ahead. Sustainability means we should not lock urban development and current expensive long-term investments of infrastructure into a pattern that ultimately will need to be abandoned. The costs of risks and risk mitigation should be balanced equitably between current and future generations.

Institutional Structures

A large and varied group of institutions and governing bodies throughout the MEC region is involved in tasks relating to coastal management (Zimmerman, 2001). Some key organizations and agencies and their functions are summarized in Table 9.2 (see also Appendix Decision-Making 1 and 2, in Rosenzweig & Solecki, 2001). Fragmentation of jurisdictions among federal, state, and local governments and inter-agency communication shortcomings hinder development of a coherent, comprehensive regional coastal management plan. Authority and responsibilities are highly specialized by function and territory. On the other hand, new plans for regional capital improvements can be designed to include measures that will reduce vulnerability to the adverse effects of SLR. Wherever plans are underway for upgrading or constructing new roadways, airport runways, or wastewater treatment plants which may already include flood protection, these need to be planned or modified to take projected SLR into consideration. The extent and effectiveness of such adaptive measures will depend on building awareness of these issues among decision makers, fostering processes of interagency interaction and collaboration, and developing common standards (Zimmerman, 2001).

A distinctive feature of the MEC project was the participation of key stakeholders throughout the assessment process. Stakeholders consisted of institutions or groups "whose activities are and will be impacted by present and future climate variability and change, and thus have a stake in being involved in research of potential climate impacts". Groups included federal, state, and local government agencies, as well as several universities (Table 9.3). Stakeholder partners, such as the U.S. Army Corps of Engineers, provided critical data, model outputs, as well as offering advice and feedback.

The Port Authority of New York and New Jersey (PANYNJ) is a major regional owner/operator of infrastructure facilities (ports, airports, bus terminals, major trans-Hudson bridges and tunnels, and — until 2001 — the World Trade Center (WTC). PANYNJ was also a very active stakeholder and participant in the MEC study. It had already undertaken several mitigation measures at its facilities, although some of these measures will be effective only for a limited time and may eventually be overtaken by SLR, and thus may need further modifications. It has built, for instance, a dike and levee

Table 9.2: Selected key institutions in the New York City Metropolitan area with a stake in coastal zone management.

Organization	Jurisdiction	Function/Authority
NYS DOS (Department of State) Div. Coastal Resources	NY	Coastal zone management planning.
NJ Office of State Planning	NJ	Coastal zone planning, land use
CT Department of Environmental Protection	CT	Issues permits to regulate development: coastal management act, tidal wetlands, structures, dredging and fill
NJ Department of Environmental Protection	NJ	Waterfront Development Law, Coastal Area Facility Review Act, Wetlands Act of 1970, Flood Hazard Area Control Act, and the Tidelands Act
NY State Department of Environmental Conservation	NY	Environmental Conservation Law Permits-Protection of Waters, Tidal Wetlands, State Water Quality Certification
U.S. Army Corps of Engineers-NY District	NY and NJ	Dredge and fill permits, shipping channels (Rivers and Harbors Act), wetlands permits (Clean Water Act), beach nourishment
Port Authority of New York and New Jersey	NY and NJ	Develops, operates, maintains Port Authority bridges, tunnels, PATH trains, port facilities, ferries, and airports
New York City Department of Transportation	NYC	Operates and maintains city-owned roads and bridges
New York State Department of Transportation	NY	Operates and maintains state-owned roads and bridges
Metropolitan Transportation Authority	Mainly NYC	Owns, manages, operates, maintains New York City subway system
New Jersey Transit	NJ–NY	Owns, manages, operates, maintains buses and trains linking northeast New Jersey with New York City
NJ Office of State Planning	NJ	Infrastructure needs Assessment 2000–2020
NYC Department of Environmental Protection	NYC, upstate	Owns, manages, operates, maintains wastewater treatment plants, sewers, and associated equipment (pumps, regulators, etc.)

(*Continued*)

Table 9.2: (*Continued*)

Organization	Jurisdiction	Function/Authority
NYC Office of Emergency Management	NYC	Responds to natural and man-made disasters
NYS Emergency Management Office	NYS	Responds to natural and man-made disasters
NJ Office of Emergency Management	NJ	Responds to natural and man-made disasters
CT Office of Emergency Management	CT	Responds to natural and man-made disasters
FEMA, Region II	NYS and NJ	Responds to natural and man-made disasters; National Flood Insurance Program (NFIP).
FEMA, Region I	CT	Responds to natural and man-made disasters; National Flood Insurance Program (NFIP)

Source: Zimmerman (2001).

Table 9.3: Stakeholder partners in the MEC project/coastal zone study.

Name	Organization
Stephen Couch	U.S. Army Corps of Engineers, New York District
Bruce Swiren	Federal Emergency Management Agency, Region II
Christopher Zeppie	Port Authority of New York and New Jersey
John T. Tanacredi[a]	National Park Service, Gateway National Recreation Area
Frederick Mushacke	New York State Department of Environmental Conservation
David Fallon[b]	New York State Department of Environmental Conservation

[a]Now at Department of Marine Sciences, Dowling College.
[b]Retired.

system around the LaGuardia Airport, one of the three major regional airports. Before the protective measure, LaGuardia Airport had been repeatedly flooded especially by nor'easter storms, as early as the 1950s. The severe nor'easter storm in 1992 flooded the PATH tunnel under the Hudson River, used by commuter trains between Hoboken (New Jersey), and Manhattan (New York). PANYNJ built floodgates at the tunnel entrance and provided other flood protection on the NJ side from which the floodwaters had entered. The flood put the PATH trains out of operation for 10 days. One of the first construction projects after the terrorist attacks on the World Trade Center on September 11, 2001, was for PANYNJ to raise at the WTC site the perimeter slurry wall to an elevation of 1 ft (about 0.3 m) above the FEMA-established 100-year flood elevation. When hurricane *Isabel* threatened the U.S. east coast in 2003, resident engineers at the WTC reconstruction site ensured that material and equipment was readily at hand to seal temporary construction entry ways through the slurry wall, should this become necessary. It turned out it was not needed. But given the rate of SLR, the voluntary 1-ft extra margin of the slurry wall above the FEMA/NFIP-set levels for the 100-year flood elevation will be erased probably before the year 2050. This important downtown Manhattan redevelopment and reconstruction project could set an example for forward-looking preventive measures that anticipate the projected, accelerated SLR and the resulting increased storm surge flood hazards.

Conclusions

The Metropolitan East Coast (MEC) vulnerability assessment and report (Rosenzweig & Solecki, 2001) for the *National Assessment of Potential Consequences of Climate Variability and Change for the United States* is an important, albeit small step toward facing the increasing storm surge flood hazards and risks for this highly urbanized and increasingly vulnerable region. Key findings include a potential regional rise in sea level between about 0.24 and 1.08 m by the year 2100, a marked increase in the storm flood recurrence frequency (i.e. shortening of recurrence periods from 100-years to as little as 4 years for the fastest SLR scenario), and 4–10 times greater beach erosion by the 2080s, as compared to late 20th century rates.

Since many elements of the regional transportation and other infrastructure and built assets lie at elevations of 2–6 m above present sea level they are exposed to current and projected future coastal storm surge risks from tropical and extra-tropical cyclones. Major and wide-spread damage occurring during high-impact, low-frequency storm surges can measure in the tens of billions, and for the most severe storm scenarios could well exceed U.S. $100–200 billion. Such losses are expected to have significant repercussions for the regional economy. The economic effects may last a decade or more. The rate of storm-surge risk (annualized long-term average of future losses) is expected on average to increase three-fold during the current century, just from SLR alone, not counting that new vulnerable assets may be added in hazardous locations, and that climate change may increase the storm intensities and frequencies.

To protect the current and future assets of the metropolitan New York City region, and to enhance its reputation as a safe and attractive global center for business, trade, culture education and diplomacy, a coherent long-term plan for coastal risk management and adaptation needs to be implemented. The singular and limited efforts of a few institutions to address the coastal risks and adaptation options are inadequate for the task facing the region. Evacuation planning is reasonably well advanced but largely untested for truly severe scenarios. Turning water front property and former piers, for example on the west-side of Manhattan facing the Hudson River into park and recreational facilities, are encouraging protective measures. But these positive measures do not address the greatest risks, especially for the very vulnerable low-lying infrastructure virtually unmitigated. The institutional authority and responsibility is often narrowly defined, fragmented and hampered by a historically based, divided, yet often overlapping set of jurisdictions and public functions. This institutional disposition tends to foster bureaucracies that are more interested to protect their own survival and economic basis than that of the common good and public safety. The periodically low interest at various levels of the federal government in the causes and effects of climate change does not release the local and state governments from their responsibilities to act on behalf of the long-term safety interests of the regional and local public. Public safety is largely a state and local function in the U.S.

Perhaps, the scientific and engineering professional communities will need to define the current and increasing risks more clearly, convey them in unison and hence more forcefully, without shying away from pointing out the inherent uncertainties. But inherent uncertainties cannot be used as an excuse by decision makers for public inaction especially when mean trends and observable facts and analyses are on average clearly pointing to ever-increasing predictable risks.

A state-governor-appointed task force at least involving New York and New Jersey (but also perhaps Connecticut) may be a much-needed option, with proper authority and input especially from the New York City government, to develop overarching regional plans and priorities. In addition, New York City needs its own long-term master plan to address its internal coastal and waterfront storm-surge risks. But this City effort needs full integration with the regional state plans. Without the political will and vision, supported by sound science and engineering, the region will face ever-increasing coastal risks, and eventually inevitable (yet partly avoidable) catastrophic losses.

A congressionally mandated, federally funded new study to be released soon (NIBS-MMC, 2005) assesses how beneficial the return from every dollar invested in mitigation of natural hazards is, based on U.S. (FEMA) experience over the last few decades. The study reports a wide range of benefit-to-cost ratios depending on hazard, project type, location, and mitigation process. Benefit–cost ratios for virtually all assessed mitigation projects was found to be above 1, with many clustering around a 3-dollar return for every 1-dollar investment of risk mitigation, and some substantially higher returns. One wonders when decision makers in the communities, and the public at large, will catch up with sound business practice as the basis for managing coastal risks and adaptation to a dynamically changing environment. The ultimate goal of preventive and adaptive actions is to save lives and make human activity sustainable. This truism becomes especially important for regions, like the MEC, where populations and assets are concentrated in large vulnerable coastal cities.

References

ACOE/FEMA/NWS. (1995). *Metro New York hurricane transportation study*. Interim Report. US Army Corps of Engineers, Federal Emergency Management Agency, National Weather Service, NY and NJ State Emergency Management Offices.

Anonymous. (2004). *AIR analysis concludes 2004 hurricane season is not so unusual*. Available online- http://www.air-worldwide.com/_public/html/newsitem.asp?ID=634 (last accessed 2nd March 2006).

Bowman, M. J., Colle, B., Flood, R., Hill, D., Wilson, R. E., Buonaiuto, F., Cheng, P., & Zheng, Y. (2004). *Hydrological feasibility of storm surge barriers to protect the metropolitan New York – New Jersey Region*. Summary Report. Stony Brook: Marine Sciences Research Center, State University of New York.

Cabanes, C., Cazanave, A., & Le Provost, C. (2001). Sea-level rise during past years determined from satellite and *in situ* observations. *Science, 40*(294), 840–842.

Church, J. A., White, N. J., Coleman, R., Lambeck, K., & Mitrovica, J. X. (2004). Estimates of the regional distribution of sea level rise over the 1950–2000 period. *Journal of Climate, 17*, 2609–2625.

Dyurgerov, M. B., & Meier, M. (2000). Twentieth century climate change: Evidence from small glaciers. *Proceedings of the National Academy of Sciences of the United States of America, 97*(4), 1406–1411.

Elsner, J. B., Jagger, T., & Niu, X. F. (2000). Changes in the rates of North Atlantic major hurricane activity during the 20th century. *Geophysical Research Letters, 27*, 1743–1746.

FEMA. (1997). *Answers to questions about the National Flood Insurance Program*. Federal Emergency Management Agency. Washington, DC: U.S. Government Printing Office. See also- http://www.fema.gov/nfip/qanda.shtm and http://www.fema.gov/nfip/intnfip.shtm (last accessed 2nd March 2006).

FEMA. (2005). *Software program HAZUS-MH MR1 for estimating potential losses from multi-hazard disasters.* It has an earthquake, flood and wind module. Periodical upgrades; current release February 2005. The Federal Emergency Management Agency. Available online-http://www.fema.gov/hazus (last accessed 2nd March 2006).

Goldenberg, S. B., Landsea, C. W., Mestas Nuñez, A. M., & Gray, W. M. (2001). The recent increase in Atlantic hurricane activity: Causes and implications. *Science, 293*, 474–479.

Gornitz, V. (2001). Sea level rise and coasts. In: C. Rosenzweig, & W. D. Solecki (Eds), *Climate change and a global city: The potential consequences of climate variability and change-metro east coast.* Report for the U.S. Global Change Program, National Assessment of the Potential Consequences of Climate Variability and Change for the United States. New York: Columbia Earth Institute.

Gornitz, V., Couch, S., & Hartig, E. K. (2002). Impacts of sea level rise in the New York City metropolitan area. *Global and Planet Change, 32*, 61–88.

Gregory, J. M., Huybrechts, P., & Raper, S. C. B. (2004). Threatened loss of the Greenland ice-sheet. *Nature, 428*, 616.

Hartig, E. K., Gornitz, V., Kolker, A., Mushacke, F., & Fallon, D. (2002). Anthropogenic and climate-change impacts on salt marshes of Jamaica Bay, New York City. *Wetlands, 22*, 71–89.

Houghton, J. T., Ding, Y., Griggs, D. J., Noguer, M., van der Linden, P. J., Day, X., Maskell, K., & Johnson, C. A. (Eds). (2001). *Climate change 2001: The scientific basis*. Intergovernmental Panel on Climate Change, Cambridge: Cambridge University Press.

Jacob, K. (2001). Infrastructure. In: C. Rosenzweig, & W. D. Solecki (Eds), *Climate change and a global city: The potential consequences of climate variability and change – metro east coast.* Report for the U.S. Global Change Program, National Assessment of the Potential Consequences of Climate Variability and Change for the United States. New York: Columbia Earth Institute.

Leatherman, S. P., & Allan, J. R. (Eds). (1985). *Geomorphic analysis of south shore of Long Island Barriers, New York*. New York: U.S. Army Corps of Engineers.

Leatherman, S. P., Zhang, K., & Douglas, B. C. (2000). Sea level rise shown to drive coastal erosion. *EOS, 81*, 55–57.

Levitus, S., Antonov, J. I., Wang, J., Delworth, T. L., Dixon, K. W., & Broccoli, A. J. (2001). Anthropogenic warming of the earth's climate system. *Science, 292*, 267–274.

McCarthy, J. J., Osvaldo, F., Canziana, N. A., Dokken, D. J., & White, K. S. (Eds). (2001). *Climate change 2001: Impacts, adaptation and vulnerability*. Contribution of the Working Group 11 to the Third Assessment Report of the Intergovernmental Panel on Climate Change (IPCC). Cambridge: Cambridge University Press.

Mileti, D. (1999). *Disasters by design. A reassessment of natural hazards in the United States*. Washington, DC: Joseph Henry Press.

NIBS-MMC. (2005). *An independent study to assess the future savings from mitigation activities*. Report to Federal Emergency Management Agency and the US Congress. National Institute for

Building Sciences – Multihazard Mitigation Council. Available online-http://www.nibs.org/MMC/mmchome.html (last accessed 2nd March 2006).

Nicholls, R. J. (1995). Synthesis of vulnerability analysis studies. In: P. Beukenkamp, P. Günther, R. J. T. Klein, R. Misdrop, D. Sadacharan, & L. P. M. de Vrees (Eds), *Coastal zone management centre publication 4. Proceedings of the World Coast Conference 1993;* Noordwijk, The Netherlands, 1–5 November, 1993. The Hague: National Institute for Coastal and Marine Management.

NOAA/NOS. (2005). *Sea level trends.* National Oceanic and Atmospheric Association, National Ocean Service. Available online-http://140.90.121.76/sltrends/sltrends.shtml (last accessed July 10 2005).

Peltier, W. R. (2001). Global isostatic adjustment and modern instrumental records of relative sea level history. In: B. C. Douglas, M. S. Kearney, & S. P. Leatherman (Eds), *Sea level rise – history and consequences.* San Diego, CA: Academic Press.

Rosenzweig, C., & Solecki, W. D. (2001). *Climate change and a global city: The potential consequences of climate variability and change – metro east coast.* Report for the U.S. Global Change Program, National Assessment of the Potential Consequences of Climate Variability and Change for the United States. New York: Columbia Earth Institute.

Thieler, E. R., Pilkey, O. H., Jr., Young, R. S., Bush, D. M., & Chai, F. (2000). The use of mathematical models to predict beach behavior for U.S. Coastal Engineering: A critical review. *Journal of Coastal Research, 16,* 48–70.

Thomas, R., Rignot, E., Casassa, G., Kanagaratnam, P., Acuna, C., Akins, T., Brecher, H., Frederick, E., Gogineni, P., Krabill, W., Manizade, S., Ramamoorthy, H., Rivera, A., Russell, R., Sonntag, J., Swift, R., Yungel, J., & Zwally, J. (2004). Accelerated sea-level rise from West Antarctica. *Science, 306,* 255–258.

Zhang, K., Douglas, B. C., & Leatherman, S. P. (2000). Twentieth-century storm activity along the U.S. East Coast. *Journal of Climate, 13,* 1748–1761.

Zhang, K., Douglas, B. C., & Leatherman, S. P. (2004). Global warming and coastal erosion. *Climatic Change, 64,* 41–58.

Zimmerman, R. (2001). Institutional decision-making. In: C. Rosenzweig, & W. Solecki, (Eds), *Climate change and a global city: The potential consequences of climate variability and change – metro east coast.* Report for the U.S. Global Change Program, National Assessment of the Potential Consequences of Climate Variability and Change for the United States. New York: Columbia Earth Institute.

Chapter 10

Promoting Sustainable Resilience in Coastal Andhra Pradesh

Peter Winchester, Marcel Marchand and
Edmund Penning-Rowsell

Introduction

Successful coastal management depends on reconciling the interests of all those who live in coastal areas so that no one group of people is particularly favoured or disadvantaged by decisions concerning the distribution of resources. This issue is all the more stark where most of the population in the area in question, by world standards, is highly disadvantaged. Such is the situation in much of coastal India.

Coastal Andhra Pradesh is prey to cyclones, sea flooding and drought. Although the damage and disruption they bring in their wake is relatively localised, some people suffer more from the impact of these events than others. Widespread environmental deterioration and increasing competition for resources greatly exacerbates their suffering. But the issues are not simple. The local political economy of coastal Andhra Pradesh tends to govern peoples' lives here — as in most of India — and their asset and resource base largely determines their vulnerability to the impact of cyclones and other climate extremes, as well as their capacity to recover from them. However, their resilience — their capacity to adapt to these events — is not wholly governed by economic factors but also by their attitudes to uncertainty and opportunities.

In order to understand these issues we need to appreciate that there are a number of ways in which different sections of the coastal population adapt to natural or human-induced vicissitudes. We also need to know that there are large socio-economic differences within and between apparently homogeneous groups living in the same places. In order to solve any of the problems inherent in coastal management in Andhra Pradesh, and possibly elsewhere, we need to know the causes of these differences and why they persist.

The findings in this chapter are based on research over a period of 25 years in one locality of coastal Andhra Pradesh (Winchester, 2000) and from a single broader analysis of 20 areas along the 1000 km state coastline (Winchester & Penning-Rowsell, 2001). Results

Managing Coastal Vulnerability
Copyright © 2007 by Elsevier Ltd.
All rights of reproduction in any form reserved.
ISBN: 0-08-044703-1

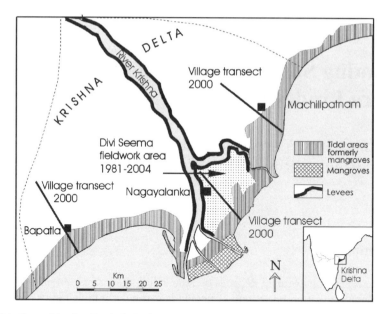

Figure 10.1: Coastal Andhra Pradesh, India, showing the Divi Seema study area and the village survey transects.

(see Figure 10.1), published by Winchester (1992, 2000), were derived from fieldwork in Divi Seema from 1981 to 2004 from a sample survey of 202 households. Results in Winchester and Penning-Rowsell (2001) were derived from fieldwork in 10 transects along the coast of Andhra Pradesh in 2000 from a sample survey of 1000 households. Each transect was 20 km deep and contained around 100 households in three or four villages selected as the most representative of the socio-economy of each transect.

The perspective from this intensive longitudinal research and extensive regional analysis is, first, that we do not fall into the trap of suggesting simple answers to complex problems, but also, second, we know that any search for solutions starts with an assessment of vulnerability and its root causes within the socio-economic and environmental context in which this vulnerability occurs.

The evidence from this case study suggests that improving the socio-economic environment, particularly through measures that build on peoples' innate capacity to adapt, may be the best way to mitigate both environmental deterioration and the impacts of the area's extreme natural events.

Setting the Context of Vulnerability in Coastal Andhra Pradesh

Coastal Andhra Pradesh is generally very poor, except the Krishna and Godavari deltas. The Government of Andhra Pradesh attributes the historically low economic growth here, and the relative backwardness of their populations, to the area's inaccessibility, its low-lying nature and the episodic but continuous impacts of cyclones and sea flooding. It sees its first development priority as the "management" of the physical environment and consequently allocates a considerable amount of its coastal development budget to raising standards of flood and cyclone protection and to improving the physical environment (Government of Andhra Pradesh, 1977, 1979).

A parallel view, held by some in government but mainly in the NGO world, is that the economic and cultural backwardness of the coastal areas would be reduced and gradually eliminated if the innate skills of the native populations to adapt to difficult circumstances could be harnessed to improving their own environments, through self-management. This, however, would need much better management of the overall socio-economic environment than the one that exists today.

Approximately 14 million people live within 20 km of the 1000 km coastline of Andhra Pradesh (Government of Andhra Pradesh, 2001). Some stretches of the coastline are more prone to cyclones and sea flooding than others but the Government considers that the populations within the entire 20 km band are more or less equally vulnerable to loss of life and damages on account of their exposure to cyclones and river and sea flooding. Except in very rare occasions (1977, 1979, 1996), the coastal populations are more vulnerable to loss of life and damages from cyclones and river flooding than the inland populations. However, within the 20 km band the loss of life and damages vary widely depending on variations in topography, cropping patterns, planting regimes, the agricultural calendar, the extent of natural protection, and, other factors that we discuss later.

Cyclones are dramatic events. Loss of life and widespread damages remind people that humans are almost powerless in the face of such huge natural forces and only governments have the will and the resources to combat them. Governments' perceptions about how to combat such forces are therefore generally accepted by the populace at large.

Cyclones and floods occur seasonally but irregularly (Table 10.1). Some cyclones and floods kill hundreds of people and thousands of animals over areas up to 500 km^2, while others will damage an area no larger than 25 km^2 and kill only a handful of people and a hundred animals. Storm surges are the most dangerous in terms of loss of life and accurate forecasting and timely warnings are the best defence (Delft Hydraulics, 1999–2003).

Table 10.1: The incidence and impact of severe and normal cyclones crossing the Andhra Pradesh coastline between 1949 and 1983 (Winchester, 1992, p. 7).

	Severe cyclone storm (high wind, heavy rain with storm surge)	Severe cyclone storm (high wind and heavy rain)	Severe cyclone (low wind and heavy rain)	Cyclone Storm (low wind and light rain)
Number of cyclones	5	4	3	2
Mean numbers of people killed	5591	833	59	17
Damage to crops in mean lakh (10,000) hectares	10.70	3.78	1.86	1.22

Table 10.2: Density of population per square km in districts of Andhra Pradesh.

Coastal districts	Density	Non-coastal districts	Density
Srikakulam	434	Chittoor	247
Vizianagaram	343	Anantapur	190
Visakapatnam	342	Kurnool	200
East Godavari	453	Mahbubnagar	190
West Godavari	491	Ranga Reddy	479
Krishna	479	Hyderabad	17,632
Guntur	391	Mezak	275
Prakasam	173	Nizamabad	294
Nellore	204	Adilabad	154
		Karimnagar	296
		Warangal	252
		Khammam	160
		Nalgonda	228
Average density of coastal districts	368	Average density of non-coastal districts (except Hyderabad)	247

Source: Districts at a glance in Andhra Pradesh 2003. Directorate of Economics and Statistics, Andhra Pradesh, Hyderabad

The cyclone prone areas of coastal Andhra Pradesh are thickly populated (Table 10.2). At the beginning of the 20th century the spread of canal and tank irrigation into the coastal areas (particularly in the Krishna and Godavari deltas) attracted an influx of cultivators from inland areas who displaced the previous inhabitants of a low-density mixed population, pushing them towards the coast into the lower lying areas of marginal agricultural value. As a result of the rapid population growth (200% in 50 years) about 70% of the natural mangrove forests in the coastal areas (currently 30,000 ha) were cut down and the protection they offered against cyclones and sea flooding all but vanished. Other forests also disappeared (EPTRI, 1999).

Today, much of the area once occupied by mangroves and forests is under large-scale *paddi* and cash crop agriculture and aquaculture, or are degraded wastelands resulting from overgrazing and tree cutting. More than 80% of the rural households in the coastal areas depend on firewood for cooking yet there is currently estimated a deficit of 70% of firewood production (Government of Andhra Pradesh, 1991; Marchand & Mulder, 2003). But cyclones and population density are not the only issues here. Water supply is a critical issue for 20% of the inhabitants of the coastal settlements who are totally dependent on poor quality, disease ridden, groundwater or surface drinking water, for example brackish water or water with an excessive natural fluoride concentration. The increase in groundwater use for domestic and agricultural purposes has also resulted in greatly increased salt intrusion in the groundwater aquifers in large areas of the Krishna and Godavari deltas.

Land degradation due to soil salinisation, malfunctioning irrigation systems, tree cutting and overgrazing have all eroded the ecological base of the predominantly agricultural livelihood systems of the rural population (EPTRI, 1999). In many respects, and for many people today, the conditions of everyday life in the coastal areas, including the two prosperous deltas, could be described as somewhat "hostile" (Sanjeeva, 1999).

But the activity that has most affected the rural economy of the coastal areas in the last 20 years has been the conversion of wastelands and mangroves into brackish water prawn aquaculture, and lakes and paddy fields into fresh water prawn aquaculture (Swaminathan Foundation, 1998). After spectacular growth in the late 1980s and steady expansion in the 1990s the aquaculture industry more or less came to a halt in the early 2000s due to persistent outbreaks of viral infection (Singh, 2001). The introduction of aquaculture was contentious from the start. Its opponents argued that it would lead to a loss of mangrove forests and mainly marginal but still productive agricultural land, an increase in water pollution and salinity in the groundwater, loss of jobs and the un-sustainability of the crop production itself. Those who argued in favour of aquaculture development pointed to the additional employment created in the marginally productive agricultural areas (Murthy, 1999).

The prawn "boom" indeed boosted the economy of the coastal region estimated at rupees 123 crores in 2000 (Singh, 2001), making Andhra Pradesh the largest producer of brackish water prawns in India. But its impact brought mixed blessings to local livelihoods. In the early days many small landholders and fishermen made relatively large profits but very many went bankrupt. However, landless labourers continued to make profits by catching prawn "seedlings" in the wild and then breeding them (a relatively simple bucket operation) and this activity played some part in boosting local incomes (Thakur, Reddy, & Prakash, 1997).

On the whole the opponents of aquaculture were correct in their prognostications but not all the most noticeable features of environmental deterioration can be attributed to aquaculture. The most widely recorded direct consequences of its introduction have been the widespread decline in drinking water quality resulting from increased groundwater pollution from flushing out prawn tanks, and a further reduction in "free" grazing land for livestock and subsequent fodder availability resulting from the taking of "common" lands for prawn tanks. This may have compounded the increasing shortage of fuel wood (mainly the result of population pressure), and the further reduction of protection against cyclones by clearing the remaining accessible areas of mangrove forests.

Thus, here there are many other intricate and complex relationships between the environment — its use and abuse — and the livelihood of rural people, and their interaction is bound to affect their vulnerability to extreme natural events (Delft Hydraulics, 1999–2003). The arena in which decisions are made concerning the allocation and distribution of resources is referred to throughout this chapter as the "political economy", which, it will be argued, is the key factor in determining the vulnerability to, and recovery from, damaging events of a community, households or individual.

The Allocation and Distribution of Resources – A Crucial Factor in Social Vulnerability

The Government of Andhra Pradesh has made it generally safer to live in the coastal areas and road communication is now easier than it was 20 years ago. All weather roads now connect 65% of all villages in coastal Andhra with each other (Marchand & Mulder, 2003) and the government has also instituted wide ranging economic programmes such as DWCRA, NABARD, VELEGU (Government of Andhra Pradesh, 1999). Nevertheless, 65% of the population are cultivators of less than two acres of land and landless labourers, and these people are generally not benefiting from the increasing overall prosperity (Winchester & Penning-Rowsell, 2001).

Most of these people still have difficulty in getting access to cheap credit or affordable health care or the costs associated with basic education (SADA, 1988). This is because in most places the Government of Andhra Pradesh has yet to solve the problems associated the allocation and distribution of resources within the local communities (Winchester, 2000).

The allocation and distribution of resources is by its nature political, and functions at two separate but closely linked political levels — the State level and the District/local level. Decisions taken at both levels reflect political realities rather than actual problems "on the ground". Where decisions are made with the emotional backdrop of loss of life and damage from cyclones and floods the decisions become much more complicated and doubly political and there is always competition for relief resources from other parts of the State, particularly from the drought-prone inland districts that do not have the insurance of canal-fed irrigation.

Decisions at State level about the best allocation of resources are governed by the wider political economy: what is of concern elsewhere in the State, the State government's relationships with the other districts and with the Government of India. At the District and local level, decisions are governed by a simple local political economy in which the large landowners, village officials and elders obtain government resources and divert them to their families and political friends in what they consider to be "best" allocation of resources because up until recently the monitoring structures have been weak. There are many government anti-poverty initiatives in place to reach the most needy (the most vulnerable) in Andhra Pradesh but prevailing pressures within the local political economies continue to make life difficult for the (minor) government officials who administer them.

The human resources and asset resource base of a household is generally considered (Winchester, 2000) to be a reliable indicator of its susceptibility to, and ability to recover from, some extreme event or other (its vulnerability) but, not necessarily its capacity to adapt to them. Table 10.3 illustrates the recovery trajectory of a sample of households over 20 years, measured by the three key assets of a rural household — land, plough animals and other animals. In that period there have been 14 cyclones, six floods and nine extremes of climatic variation (Delft Hydraulics, 1999–2003). Despite these damaging events it would appear that those households with a wide asset base before the great cyclone of 1977 have prospered since.

The differences in land ownership over this period (Table 10.3) can be explained by the rewards and perils associated with spreading or not spreading risks and are almost entirely attributable to the prawn boom (actively encouraged by Government) that collapsed (owing to viral attack) resulting in many bankruptcies among the small landowners. The small landowners had no means to cover their risks (i.e. their debts), whereas the large

Table 10.3: Recovery between 1977 and 1997 in the Krishna delta (Winchester, 2000).

Economic classification (households)	Changes in land ownership (ha)	Other assets (plough teams and carts)	Animals
Rich ($n=16$)	+1.5 (range = 2.0–3.5)	−2.2 (range = 3.4–5.6)	+3.7 (range = 2.5–6.2)
Medium ($n=60$)	−0.25 (range = 0.75–0.5)	+0.7 (range = 0.5–1.2)	+1.2 (range = 1.3–2.8)
Poor ($n=34$)	0.0	0.0	+0.1 (range = 0.1–0.2)

landowners had reserve land on which to cultivate a low-risk crop (*paddi*) and could afford to "sit out the crisis". The land previously owned by many small landowners passed, via the banks, into the hands of the large (rich) landowners.

The differences in the ownership of recoverable assets — plough teams, carts and animals (Table 10.3) can also be explained by the same process as above. After cyclones and floods the government compensates those who can prove their losses but compensation takes time to administer and further differentiates between those who can afford to wait the necessary 6 months on average and those who can or cannot pay the necessary "handling charges", i.e. spreading their risks by "servicing" the political economy.

The most successful households (the rich) are large and multi-occupational and most of their family members are educated. They live in thick walled well-built houses on relatively high ground. They have a network of family members who live nearby or in an inland town or city well outside cyclone range, any of whom who can provide financial assistance after a catastrophic flood or cyclone. These households have a variety of income sources and, over the years, have accrued enough surplus money to make them resilient to shocks such as from cyclones, mainly from assets and money lending. They can afford to pay to those who control the local economy the necessary "handling charges" to ensure first pick of the resources available (irrigation water, cheap credit, subsidies, etc.). These households are therefore relatively invulnerable to any shock, from extremes of natural phenomena to "everyday" events such as accidents, ill health, escalating dowry prices or bribe rates. Almost the opposite is true in all respects for the least well off, who can barely help each other even after some natural calamity or other.

Field observations (Table 10.4) indicate that although all strata of society are aware that resources have been directed into improving certain aspects of the physical and economic environments in the last 20 years there is a very marked difference in perception as to where resources should still be targeted. Typically, the poor (households) are most concerned with aspects of developments that have the most impact on their health — unsafe drinking water and lack of sanitation (they cannot afford to be ill); they are also concerned with: decline in soil quality (they cultivate poor-quality land), water logging of land (their houses are invariably sited on low-lying encroachment land) and inadequate infrastructure (they depend a great deal on public transport). The rich are least concerned with everything except about the need for clean drinking water.

These data from our fieldwork (Winchester, 2000) may suggest that "money conquers all" and that an individual's or household's survival and prosperity in a risky climatic area are dependent on their relationship with those who control the local economy. Other data from the sub-continent in similar situations suggest that adaptability to changing circumstances and a positive attitude to uncertainties and opportunities, especially when there are clear prospects for improvement, may be as essential for survival and prosperity as a secure resource base. This is discussed below.

Promoting Resilience through Increasing the Capacity to Adapt

Uncertainty associated with the climate is the single-biggest factor in determining the economic well-being of most people living in the rural coastal areas of Andhra Pradesh

Table 10.4: The percentage of households in 2001 that perceived particular aspects of development as having the most negative impacts on their lives (Winchester & Penning-Rowsell, 2001; Table 3.1, Appendix IV).

Economic ranking (households)	Unsafe drinking water	Lack of water for crops	Decline in fish catch	Decline in soil quality	Decline in fodder resources	Waterlogging of land	Inadequate infrastructure	Lack of sanitation
Rich ($n=108$)	85.7	52.9	32.5	25.6	42.3	45.1	51.3	56.0
Middle ($n=577$)	87.1	42.6	46.9	24.3	39.3	45.8	61.1	60.5
Poor ($n=335$)	86.1	55.7	56.7	53.5	48.1	69.4	83.5	83.2

because the climate determines the success or failure of agriculture and that in turn more or less governs the lives of all sections of the society.

But over many years the farmers have developed a range of adaptive measures to deal with the uncertainties associated with cultivation, the most favoured being to take advantage of two planting seasons. This insurance, however, can backfire for those in the canal irrigated tail end areas (the poorest cultivators) in drought seasons when there is insufficient water to reach the tail end lands. We have mentioned the best way to spread risk by ensuring that one receives government compensation at times of hazard by paying the necessary "handling charges" usually well beforehand; other means include using expensive hybrid cyclone-resistant rice varieties, multi-cropping (three annual crops from the same ground) and intercropping. However, all these strategies are exclusively available only to the better-off farmers.

To set these uncertainties in context we should note that the probabilities of the occurrence of cyclones and associated flooding is about the same as that of extremes in climatic variation, i.e. too much or too little rain at the times of planting or harvesting (Table 10.5).

But the situation is far from static. Since the late 1970s the Government of Andhra Pradesh has greatly improved and expanded the physical infrastructure in the coastal areas in order to attract private investment and improve the safety and protection of the coastal populations. The most visible signs of physical infrastructure are the greatly expanded road network and improved drainage systems, some sea walls and flood embankments, and cyclone shelters and cyclone resistant housing. The greatest improvement to the safety of the populations has been highly sophisticated and accurate forecasting and warning programmes, pre- and post-disaster planning, and efficient evacuation procedures (Government of Andhra Pradesh, 1981).

Individuals and communities have also adapted. The ways that different income groups deal with risk and uncertainty provides us with information about their adaptability to circumstances and their relationship with their local economies. It is difficult to know which is the most important. The richer households (Table 10.3) deal with uncertainty by nurturing their place in the local political economy by "keeping in" with the powerful local and non-local individuals and at the same time by diversifying their assets in terms of type and location. Their adaptability and survival relies more on their connections in the significant local networks than, necessarily, their mental acuity.

On the other hand the poor deal with uncertainty by continuing to believe in Fate or God's will. Their relationship with the local political economy consists mainly of not falling foul of anyone in authority (their employers, minor government officials — particularly the police) at the same time as trying to ingratiate themselves with "powerful people", that is

Table 10.5: The incidence of cyclones and extremes of climate variation in coastal Andhra Pradesh (Winchester, 1992).

Climatic characteristic	Incidence (over 100 years)	Average return period in years
Cyclone	22	4.5
Severe cyclone	5	20.0
Severe cyclone and storm surge	3	33.0
Too much or too little water at planting	21	4.8
Too much or too little water at harvest	28	3.7

those who are higher up the "pecking order". Their survival depends less on networks than mental acuity and their capacity to adapt. Outsiders (NGOs, even government agencies) may try to reduce some of the uncertainties facing the poor by various means. But the poor are aware of the transitory nature and value of external agencies and, based on bitter experience, are suspicious of outsiders. The range of action for the poor is extremely limited and since they have only one asset (their labour) this is what they have to rely on.

On-going fieldwork in a small number of coastal villages in Andhra Pradesh briefly yields some examples of these processes (Box 10.1). Most poor people see choices in

---

Mental acuity [education]
The president of a village panchayat (council) in a "caste" village is a Harijan (formerly called an Untouchable) and left school at 10 years of age. Her strength is her ability to combine with others and see the advantages of "communal" effort and co-operation whereas the majority of the comfortably well-off caste villagers and her own Harijan community cannot. Through her efforts the whole village is now connected to the drinking water supply, for which she will gain further popularity and political credit.

Adaptability to changing circumstances
A group of women in a village on an island in the river took a group loan to lease land between the village and the mainland to grow and sell vegetables to richer people in a large town ten kilometres distant. Two years ago the river flooded and they lost all their vegetables. One woman repaid her loan whilst the others prevaricated and the group disbanded. The woman formed another savings group on her own initiative and after two years there was enough money for her to obtain a water buffalo loan and she started selling milk direct to her previous customers who paid good prices for very fresh buffalo milk. She still works as a labourer but now has three buffaloes and a thriving milk business.

Attitude to opportunities
Four years ago one woman with some education (but probably no more than 4 years) living in very small isolated hamlet (15 houses) in the middle of a thorn forest cam across a Christian NGO and persuaded her fellow villagers to convert to Christianity. Today the villagers are relatively prosperous; all the houses have been rebuilt and every household owns animals (for income). The group (illiterate and uneducated labourers) is very cohesive and has well understood that rigorous repayment of loans, a businesslike approach to group activity, and not overextending themselves has its rewards.

Attitude to risk and uncertainty (1)
Many fishermen regard a cyclone warning as a good opportunity to go fishing. The turbulence in the water and the atmospheric pressure bring many fish to the surface and the fisherman can catch more in a cyclone day than they can normally in a month. The government issues them with radios to take when they go fishing so that they can receive warnings and return safely, but the fishermen do the opposite. Some fishermen prosper but quite a few drown.

Attitude to risk and uncertainty (2)
The leader of a savings group living in a small mud house decided she could do better in terms of status and attracting larger loans if she built herself a concrete house. She raised most of the money through a government loan and some from private sources and certainly over-extended herself. Her fellow group members became suspicious of her 'private' funding and the group disbanded. Her interest charges and the repayments continued. After two years, and on her own initiative, she had formed another ten groups through whose savings she and others could raise loans - in her case a buffalo loan to help pay the ferocious interest charges. But, she was only able to form these ten groups because the new members considered she had status (and therefore credibility) with her concrete house.

Attitude to risk and uncertainty (3)
Even the rich can suffer, because they over extend themselves, through choice or custom. For example; a mandal president (a rich man) who grew the latest and most expensive hybrid cyclone resistant strain of rice that required expensive fertilisers and pesticides to spread his risks was caught out in 2004 when a cyclone earlier in the year was followed by a drought. He survived the cyclone damage but lost his expensive crop from the drought. Instead of paying for his second daughter's dowry from large profits from the cyclone resistant investment he had to sell five acres of medium fertility land elsewhere.

Box 10.1: Examples of adaptive behaviour from on-going fieldwork in a small number of coastal villages in Andhra Pradesh.

terms of penalties rather than opportunities and in most cases their capacity to make choices is determined by their low economic status and their place in the local political economy, their level of education, and, in rural areas, by caste. However, the examples also illustrate how poor people with very limited choices can adapt to their less than favourable circumstances. In all these examples (except the last) the people in question were vulnerable owing to some unfortunate event or other: for the *Harijan* president, for example, it was her social rank at birth. But in each case it was their ability to adapt to changing circumstances, their strength of character and mental acuity and opportunism that enabled them to reduce their vulnerability and enhance their capacity to cope with hazard and to recover.

Changing Opportunities: A Model of Coastal Development and Vulnerability

The Government of Andhra Pradesh needs to develop a policy of Integrated Coastal Zone Management (ICZM) within the background of prevailing socio-economic structures and processes in the coastal society as described above. They need to investigate a range of initiatives and measures best suited to reducing differential physical and socio-economic vulnerability and achieving the sustainable use of the coastal environment.

As one means of investigation, Delft Hydraulics developed a model (Baarse & Marchand, 2003) that predicts different scenarios of physical and socio-economic vulnerability, based on extrapolations, as well as illustrating reliable past and present conditions, based on fieldwork analysis. The model can calculate in economic and numerical terms the interaction between the impact of natural events at different scales of environmental characteristics and socio-economic groups, and over different time scales. The model links the socio-economic character of the coastal zone to the land uses and all related activities that generate income and is thus sensitive to both planned (crop selection) and unplanned (cyclone disaster induced) land-use changes. Calculations can be made for annual production, income, resource use and waste generation for any area in any time horizon.

The model has a spatial resolution of the administrative unit of a *mandal* — a geographical area of c. 10,000 ha — and can calculate the estimated annual incomes for groups based on private and income-generating assets. The model will show the differential economic effect of combinations of occupations, household size and family size, according to a range of impact scenarios and environmental conditions. It can calculate the socio-economic and environmental effects of cyclones, storm surges and floods, and the rates of recovery from these events for households in each income category by comparing the asset status of households in different income categories one year after a cyclone or storm surge with their pre-event asset status.

The model is useful in analysing the possible effects of government policy initiatives such as: changes in land use, introduction of new types of employment, distribution of relief funds, introduction of flood-resistant crops, further investment in the cyclone shelter and housing programmes, further expanding the road network and so on.

Perhaps, most importantly, the model will enable policy makers to have a sharper awareness of the environmental, social and economic complexities of these coastal areas.

At the same time — and possibly as important — the model can reveal which policy variables could be the most effective in reducing the vulnerability of coastal populations as a whole and in particular those who are in the lowest income categories.

The model (Figure 10.2) has six interrelated modules, each of which represents different domains of the coastal economy and its environment:

- Land Use Module (LUM): describes present and future land use and related activities (crops, labour demand, wage rates, gross revenues).
- Socio-Economic Assessment Module (SAM): calculates incomes and income distribution in specific areas taking into account the employment rate and regional income.
- Resource Demand and Waste Generation Module (RWM): calculates the use of water and energy resource, transportation needs and pollution loads.
- Environmental Assessment Module (EAM): calculates the environmental impact of resource depletion.
- Flooding Probability and Severity Module (FPM): calculates the geophysical aspects of storm surge flooding, taking into account the probabilities of particular hydraulic conditions and cyclone frequencies and the physical characteristics of the geophysical units. This module reflects the severity of storm surge flooding expressed in estimated casualties and damages inferred from flood levels, rates of flooding and duration.
- Flooding Vulnerability Module (FVM): combines information from LUM, SAM and FPM and calculates vulnerability of people and capital at risk, expected damages and recovery factors.

The model has examined a number of land use development scenarios incorporating differences in land use, water use and population growth trends for the next 20 years in the

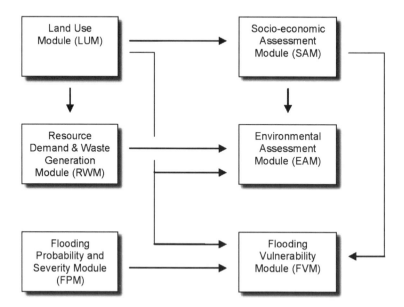

Figure 10.2: Coastal vulnerability assessment model (Delft Hydraulics, 2003).

Godavari delta (Marchand & Mulder, 2003) with the focus on the relationships between agricultural practice and employment potential and between development and environmental destruction. These scenarios were: (i) autonomous development, (ii) rice bowl, (iii) maximum land development and diversification and (iv) environmentally sound land development (Table 10.6).

The model runs suggest (Table 10.7) that the maximum land development and diversification scenario is best for employment. It is also less damaging to the environment (surface water deficit, pollution and destruction of the mangrove forests) than all the other scenarios except the environmentally sound land development scenario.

Furthermore the runs suggest that by 2020, traditional *paddi* agriculture in the Godavari delta, even with greater investment in irrigation and efficiency (the rice bowl scenario) will be unable to generate enough employment for the population which will then have increased from 7 million (2001) to 9 million (1.4% p.a). The maximum land development and diversification scenario (Table 10.7) indicates that the greatest employment potential

Table 10.6: Development scenarios (Delft Hydraulics, 2003).

	Baseline (2001)	Autonomous development (2020)	Rice bowl (2020)	Maximum land development and diversification (2020)	Environmentally sound land development (2020)
Land use	Current land use	Estimated annual trends	Priority to rice	Priority given to aquaculture (+46%) and Horticulture (+60%)	Reduction of aquaculture (−33%) priority to horticulture (+118%), reforestation and sylvo pastures
Population growth	N.A	Assumed 1.4% annual growth	Assumed 1.4% annual growth	Assumed 1.4% annual growth	Active family planning to reduce growth to 1.2% per year
Irrigation system	Current situation (70% of design cap)	No major investments (70% of design cap)	Major investments to increase irrigation capacity (85% of design cap)	No major investments (70% of design cap)	No major investment (i.e. 70% of design cap)
Ground water development	Current situation	No increase in ground water use	Increase in ground water use in 15 mandals	No increase in ground water use	No increase in ground water use

Table 10.7: Coastal vulnerability model: results for Godavari Delta (Delft Hydraulics, 2003).

	Baseline (2001)	Autonomous development (2020)	Rice bowl (2020)	Maximum land development and diversification (2020)	Environmentally sound land development (2020)
Population	7,106,000	9,379,920	9,379,920	9,379,920	8,598,260
Per capita income p.a.	Rs. 17,768	Rs. 15,649	Rs. 15,537	Rs. 16,518	Rs. 15,631
Unskilled employment rate	89%	65%	71%	70%	60%
Skilled employment rate	89%	72%	70%	78%	70%
Net agricultural profit (million Rs) (% change)	14,791 (0%)	18,553 (+25%)	15,823 (+7%)	23,938 (+62%)	19,180 (−30%)
Surface water deficit	−20.3	−16.8	−10.1	−11.1	−7.5
Surface water pollution index	7	10	11	9	9
Mangrove area (ha)	15.621	13.314	15.621	16.404	17.632
Fraction people vulnerable to financial losses	0.18	0.20	0.19	0.21	0.16

lies in improving livestock and poultry rearing, mixed cropping, and most importantly, in agricultural diversification. Agricultural diversification is likely to be most profitably directed into horticulture, sylvo-pasturalism and alternative aquaculture crops.

If these activities can be controlled (disease prevention and reduction of soil erosion) then it should be possible to have high financial yields from low-value land (Krishna, 1994). Many of these new activities would not require extra investment in irrigation and those resources could go to providing greater capacity in the power sector, better roads to markets, and more extension services, credit facilities and education in order to encourage farmers to change their practices. The new activities would also reduce environmental degradation (for example, the maximum land development and diversification, and environmentally sound land development scenarios are better at reducing the surface water deficit and the pollution index than the autonomous development scenario, Table 10.7).

Some doubts remain, however. For instance, it is not proven that the maximum land development and diversification scenario reduces physical and social vulnerability, or whether the environmentally sound land development scenario significantly reduces financial losses of the poorest in the event of cyclones and or floods. The model runs do show, however, that the rice bowl scenario (a monoculture), or the maximum land development and diversification scenario focusing on aquaculture (another monoculture) are equally risky and that crop choice and diversification are crucial. They also highlight two crucial issues that need to be resolved in the coastal areas: (i) the reversal of environmental degradation and (ii) the opportunities for more equitable growth with the introduction of new land uses.

Taking into account the projected population growth and increasing labour pool there seem to be no viable alternatives for economic development to benefit the majority of the population other than large-scale diversification in the agricultural sector. Diversification is particularly relevant in the delta areas where the resources of land and water are sustainable over the long term, but, even in these areas, the model runs cannot show definitely whether the sector can absorb the large increase in the labour supply. Both the use of resources and labour require careful planning and management and the diversification programmes need to go hand in hand with environmental protection policies that control water pollution and guarantee the continuation of designated "protected areas" such as the replanted mangrove stands (environmentally sound land development scenario).

Ways Forward?

Coastal Andhra Pradesh is poor, vulnerable to recurring cyclones and its political economy is often corrupt. The interaction between the impact of recurring cyclones and flooding, continuing destruction and degradation of the environment and unchecked population expansion would suggest that more people are likely to become more vulnerable in the coastal areas of Andhra Pradesh in the future than have been in the past.

The wide differences in the recovery from cyclones of sample households over the years between the rich and the poor and the wide differences in how they perceive their environments might suggest that the resource base of a household is the key to its place in, and, its perception of the world. But this would be a mistake. The examples of how some people *adapt* to changing and hostile circumstances suggests that those with a positive attitude to uncertainties and opportunities can survive and prosper (relatively), particularly when there are clear prospects for improvement.

The beneficial effects of the diversification of agriculture, as shown by our model results, show how this path may provide many opportunities for people to use their adaptive capacities.

174 *Peter Winchester et al.*

The critical situation that currently exists might be mitigated if government policy in coastal areas were to discourage capital intensive land users (industrial plants, factories and attendant infrastructure) that destroy their own and their surrounding environments from locating in the coastal areas, and instead encourage the proliferation of as wide a range as possible of small scale labour-intensive self-managed enterprises that depend for their survival on being in balance with their environment.

These enterprises, backed by extensive vocational education and skills training programmes would link many self-managed self help users (Savings groups, Thrift Societies and so on) with the many government anti-poverty initiatives thus forming a series of partnerships between people at grass roots and officials at District and State level (Box 10.2) thereby creating dynamic community frameworks in which the poorer people, especially,

Government Institutional	Government programmes directly accessible to people in rural areas	Community based Self Help User Group ...Structures
District Level **Zilla Parishad** The overall representative of elected officials as the coordinating body with limited executive functions at district level. Responsibility for: Distributing funds allotted to the district by Central and State Governments and raising local taxes	National Bank for Agricultural and Rural Development **(NABARD)** ←	**Samaikya** level Manager + Committee 6 Mahila Mandal leaders
Mandal level **Samiti Mandal/Panchayat** Principal representative executive body at mandal level of elected officials responsible for: Supervising the implementation of medium scale development programmes and among other things: selecting beneficiaries for anti poverty and employment programs; setting up and supervising the activities of the Mahila Mandals (Women Groups), employment and produce cooperatives, cottage industries, child and women's welfare institutions	Society for Elimination of Poverty **(Velegu)** ← District Rural Development Agency **(DRDA)**	**Mahila Mandal** level Manager + Committee 6 Self Help Group leaders
Village level **Gram Panchayat** Committees Gram Sabha of elected villagers representing one village, group of villages or groups of villagers within villages responsible for: Sanitation and waste disposal; maternity and child welfare facilities and services; supply of safe drinking water; access to agricultural inputs; construction and maintenance of roads, drains, bunds and bridges, monitoring fair price shops, propagation of family planning, maintaining postal services; promoting best communal practice of forestry and equitable access to forestry products, animal husbandry, dairy and cooperative farming and minor irrigation	Mother and Child Welfare Association **(MCA)** ← Development of Women and Children in Rural Areas **(DWCRA)**	**Self Help User Groups** Group leaders Groups of 10

Box 10.2: Community frameworks: Government institutions and community-based self-help user groups.

but not exclusively, can create economically less damaging bases for making choices and facing uncertainties (Reddy, Vinod, Chitoor, & Chitoor, 1999). But the success of this type of strategy is by no means guaranteed, and if not successful it will be the poor who suffer most and the relatively rich who will remain the more insulated from the often hostile environment in which they live.

References

Baarse, G., & Marchand, M. (2003). *Expert decision support system ICZM.* APCHMP Draft Functional Design Report. Delft Hydraulics.

Delft Hydraulics. (1999–2003). *Andhra Pradesh cyclone mitigation project (APCHMP).* Reports (1) Decision Support System for ICZM in Andhra Pradesh; (2) Framework Plan for ICZM in Andhra Pradesh; (3) Strategic Action Plan for ICZM in Andhra Pradesh. Delft Hydraulics.

EPTRI. (1999). *The state of the environment of Andhra Pradesh.* Environment Protection Training and Research Institute, Hyderabad: Government Central Press.

Government of Andhra Pradesh. (1977). *Statement on cyclone and tidal wave on 19th November 1977 by Sri Vengal Rao.* Chief Minister of Andhra Pradesh, Hyderabad: Government Secretariat Press.

Government of Andhra Pradesh. (1979). *Note on the recent cyclone by Dr Channa Reddy.* Chief Minister of Andhra Pradesh, Hyderabad: Government Secretariat Press.

Government of Andhra Pradesh. (1981). *Cyclone contingency plan of action.* Revenue Department, Hyderabad: Government Central Press.

Government of Andhra Pradesh. (1991). *Population census data.* Directorate of Census Operations. Form H-cooking fuel types. Hyderabad: Government Secretariat Press.

Government of Andhra Pradesh. (1999). *An economic survey – 1999–2000.* Finance and Planning Department, Hyderabad: Government Central Press.

Government of Andhra Pradesh. (2001). *Population census data.* Hyderabad: Government Secretariat Press.

Krishna, V. (1994). Andhra Pradesh, Indian east coast. In: E. Holmgrem (Ed.), *Environmental assessment for the Bay of Bengal Region. Report 67.* Swedmar/Bay of Bengal Programme.

Marchand, M., & Mulder, P. (2003). *Draft resource management plan for the Godavari Delta region in Andhra Pradesh.* APCHMP Technical Report. No 13, ICZM in Andhra Pradesh. Delft: Delft Hydraulics.

Murthy, D. S. (1999). Fisheries development in Andhra Pradesh. *Fishing Chimes, 19*(9), pp. 12–14.

Reddy, A. V. S., Vinod, K., Chitoor, S., & Chitoor, M. (1999). *Cyclone in Andhra Pradesh.* Hyderabad: Guntakil Press.

SADA. (1988). *Master plan for the development of Andhra Pradesh shore areas: Volumes 1 and 2.* Shore Area Development Authority, Finance and Planning Department, Hyderabad: Government Central Press.

Sanjeeva, R. P. J. (1999). North of Chennai, India: Developmental impacts on coastal ecosystems. *The Indian Geographical Journal, 74*(1), pp. 1–11.

Singh, D. (2001). *Fisheries and aquaculture in Andhra Pradesh.* APCHMP Technical Report. No 7. ICZM in Andhra Pradesh. Delft: Delft Hydraulics.

Swaminathan Foundation. (1998). *Atlas of mangroves on the Indian east coast.* Chennai, Tamil Nadu: Swaminathan Foundation.

Thakur, N. K., Reddy, A. K., & Prakash, C. (1997). Whether prawn farming is polluting the environment – a scientific appraisal. *Fishing Chimes, 17*(4), pp. 28–30.

Winchester, P. J. (1992). *Power, choice and vulnerability: A case study in disaster mismanagement in South India*. London: James and James.

Winchester, P. J. (2000). Cyclone mitigation and resource allocation. *Disasters, 24*(1), 18–37.

Winchester, P. J., & Penning-Rowsell, E. (2001). *Socio-economy of the coastal areas and cyclone vulnerability*. APCHMP Technical Report No 2, ICZM in Andhra Pradesh. Delft Hydraulics.

Chapter 11

Reducing the Vulnerability of Natural Coastal Systems: A UK Perspective

Julian Orford, John Pethick and Loraine McFadden

Introduction

At the start of a new century there is often an expressed perspective to consider the pressures ahead and to reflect on the instruments at hand to deal with these pressures. However, this contemporary concern for the future should not prevent us from reflecting on known pressures of the moment that still have to be addressed. For coastal scientists, contemporary coastal pressures have to be considered not only in the context of physically changing coastal environments, but also in the context of conflicting cultural demands of societies that are accelerating in pursuit of life styles associated with shorelines for social and economic purposes.

In this respect, vulnerability is a pertinent instrument by which we may consider where a coastal society has reached in its attempt to adjust its living functionality to the physical demands of the shoreline. Although physical coastal vulnerability is often associated with coastal sensitivity (Pethick & Crooks, 2000), it is strategic to recognise that anthropogenic alterations of coastlines in some developed nations now makes the issue of coastal sensitivity redundant, when sensitivity is measured in terms of other than human occupancy and functionality. A UK perspective on coastal vulnerability also needs to reflect on how nearly two centuries of active protective intervention in the shoreline has induced different types of coastal response from those of the original natural systems of a millennium ago.

It is an oft-asked question as to 'the state of coastal vulnerability', but though much debate surrounds the concept of vulnerability in human and ecological terms (e.g., Adger, 2000; Walker et al., 2004; Adger et al., 2005), there is a limited debate as to how this concept can be applied to the physical state of the coastal zone as reflected in the morphological and sediment dynamics of the system (McFadden, this volume). More importantly for the future, discussion is very limited as to how vulnerability can be managed to ensure a sustainable shoreline future. This discussion is concerned as to how such physical vulnerability (PV) might be specified and how such a state of vulnerability might be worsened, or more controversially, coastal vulnerability might be lessened. To do the latter, we need to understand

Managing Coastal Vulnerability
Copyright © 2007 by Elsevier Ltd.
All rights of reproduction in any form reserved.
ISBN: 0-08-044703-1

how the former may occur and draw lessons from that. This situates the debate into an important issue, that coastal PV can be influenced, both to worsen and feasibly to better it, by anthropogenic actions.

Physical Vulnerability Composition and Change

We regard PV as a mesoscaled time-based assessment (annual to decadal scale) by which coastal change occurs as a positive feedback response to coastal forcing. Hence vulnerability is relative to the potential of coastal change, where such coastal change should be observed by domain changes in mesoscale coastal behaviour, rather than day-to-day rhythmic variation around a quasi-equilibrium structure (often due to negative feedback prevalence). The issue of whether PV relates to the sustainability of the coast, allows a further layer of analysis whereby a value-laden PV tag as *increasing* or *decreasing* is identified. This usually means that the state of change is to be considered in the context of some other coastal function that is likely to be anthropogenically referenced, for example decreasing vulnerability implies increasing stability of the coastal environment in terms of position, structure and function. Decreasing PV is usually associated with a positive outcome for humans whereby coastal stability is equated with support of human functionality, whereas increasing PV identifies destabilised coastal environments and degraded human functionality. This notation is irrespective of whether PV changes carries any physical implications for the shoreline. As PV is difficult to resolve given this human calibration, it is potentially more illuminating to consider what actually drives PV.

If increasing PV reflects processes which increase the tendency for the coastal system to move in the direction of change, while decreasing PV is occasioned by the dominance of negative feedback mechanisms which resist change, then we can see PV as the balance between two conditional sets or defining modes:

1. Susceptibility: the forcing potential in the coastal system for positive feedback to predominate. Generally susceptibility would be established as a regional statement, although specific local physical transformations may occur by which susceptibility could be enhanced or diminished.
2. Resilience: the coastal system's capacity to return to a base state after exposure to natural conditions. Resilience reflects the dominance of negative feedback in the coastal response to forcing. Although resilience is a local response, any emphasis on mesoscale coastal behaviour as required for UK shoreline planning purposes (Burgess et al., 2002) identifies that such resilience is best identified at the level of specific geomorphic environment.

Table 11.1 indicates the ways in which PV status changes to reflect the relative balance of susceptibility to resilience within the system. At this stage only the sense of the state change can be identified, so when positive changes in both susceptibility and resilience occur, then PV tends to be constant: likewise PV tends to be constant when both susceptibility and resilience decrease. PV will only change when the defining modes alter contrary to each other: PV will increase with rising susceptibility and constant or reducing resilience, while PV will decrease as increasing resilience counters constant or reducing susceptibility.

This analytical approach leads us to examine a major concern related to anthropogenic intervention: how do we decrease PV? From a human perspective, the likelihood of

Table 11.1: (Un)changing coastal vulnerability (V) through changing (Δ) susceptibility and resilience. Relative increases (+) or decreases (−) in susceptibility and resilience will cause shifts in vulnerability.

		Δ Susceptibility	
		+	−
Δ Resilience	+	Unchanging V	Decreasing V
	−	Increasing V	Unchanging V

increasing PV by decreasing susceptibility is highly unlikely. The future scenarios identified by IPCC (Houghton et al., 2001) are indicative of increased coastal forcing via rising relative sea levels, with potential increased storminess leading to decreasing return periods of extreme water levels. Again while the scales of change are uncertain, the positive senses of change related to forcing (generating increasing PV) are undeniable. The probable way is to effect changes to vulnerability through changing resilience. This may be achieved by focusing on two human interventionist approaches:

1. By bolstering up the system's defensive capability to withstand change, usually through engineered coastal defences. Although this may be viewed as increasing resistance in the system (Klein & Nicholls, 1999), we still regard this as adding resilience to the coastal system in that it potentially adds capacity to withstand the effects of positive feedback trends, i.e., resilience. However, this approach is now viewed as controversial as the mood to build our way out of coastal problems is no longer acceptable to UK central government, albeit any grassroots desire for a quick engineering fix to 'hold-the-line'.
2. To promote a sufficiency of coastal sediment required to sustain contemporary coastal resilience, if not promoting an excess and storage thereof of sediment to be available for future morpho-sedimentary developments. This is a new approach to coastal behaviour and can best be described as 'sediment husbandry' by which sediment volumes are retained within the coastal zone and not lost to offshore sinks.

Both of these methodologies act as controls on PV, but it is important to recognise that functioning coastal systems can also move through natural processes towards stable configurations that effectively diminish PV. Figure 11.1a identifies the general trajectory of coastal forms as susceptibility increases and resilience decreases as a consequence. It is possible for the trajectory path to be moved to a position of accelerated decreasing resilience, by the removal of sediment volume from the shoreline. Therefore the rate of sediment volume loss to the coastal system is a key control on resilience. This will only go so far, as the domain of the coastal form may restructure itself into a more stable configuration in the light of sediment deficiency. This attribute of change is a measure of the self-organising capability of the coastal system (Forbes et al., 1995). The most frequently observed coastal switch, in the context of mobilised coarse-clastic sediment, is the movement from a drift-aligned coastal morphology (Figure 11.1b) to that of swash-aligned morphology (Figure 11.1c). In the transition the relative loss of longshore sediment is minimised by the movement of the morphology to support a net zero-longshore transport rate. The capacity of the system's increasing susceptibility might generate greater swash elevation sufficient for barrier overwash that will roll a

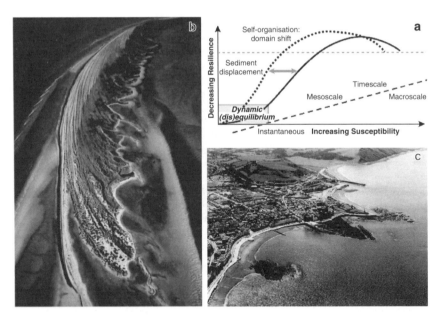

Figure 11.1: (a) The shift from sediment-rich to sediment-deficit systems engenders a loss of resilience that is accelerated by any increase in susceptibility. For some coastal systems there is a potential threshold where self-organisation abilities allow a stabilisation, or gain in resilience for the coastal form. Such a situation is indicated when drift-aligned beach/barrier systems (b) transform into swash-aligned position (c) due to sediment deficiency. The susceptibility–resilience relationship is offset (worsened) by any loss of sediment from the beach system. (b) is an oblique view of the drift dominated spit developing westwards from the coastal exit of the river Spey (Inverness, Scotland. Photograph origin is unknown). (c) is an oblique aerial view looking south over the swash-aligned and segmented beaches of Aberystwyth (Ceridigion, west Wales: photograph courtesy of Ceredigion Country Council).

coastal barrier back into a swash alignment. This domain change stabilises and even raises the resilience of the barrier to further susceptibility changes. One would expect such natural changes to take place as a function of sediment source–sink variation over hundreds, rather than thousands of years. At that latter time scale, it is likely that sea-level changes would be effective in altering the coastal system through engaging new terrestrial sediment inputs (coastal erosion) to counter diminishing resilience through sediment losses. It is arguable that human intervention in coastal sediment pathways could speed this transition rate.

Changing UK Coastal Vulnerability: Anthropogenic Approaches to Changing Resilience

There are two traditional approaches to human-generated resilience change to coastal systems: coastal defences and sea-flood defences (Table 11.2). The latent function of both

Table 11.2: Varying manifest and latent coastal resilience outcomes based on engineered protection.

	Manifest function for human resilience	**Latent function for coastal resilience**
Coastal defences	Preventing terrestrial/infrastructural/built environment losses by deflecting/absorbing kinetic power of wave activity	1. Loss of beach face sediment via scouring and/or deflected offshore pathways
Sea-flood defences	Preventing flood losses by prevention of marine incursions	2. Loss of storm deposited sediment to back beach areas

of these approaches are now viewed as causing long-term damages to coastal resilience. However viewed from the human valuation they can be seen as accelerating resilience, even if it is a revised definition of resilience change, that of human occupancy resilience, rather than coastal system's resilience. It might be argued that such engineering is resistance rather than coastal resilience *per se*, but when viewed as an extreme end case of the definition that engineering allows the coastal system to return to its original position prior to the forcing, then engineering must be defined as adding resilience as the coastal system returns to its original position, by not changing at all under forcing.

In the UK the use of coastal defences has been a long and well-honoured tradition, bedded into a 19th-century Victorian perspective that mechanistic Britain could tame the wilds of nature. Even by the start of the 20th century, Britain recognised coastal erosion as a major ongoing issue, but still regarded the use of engineered coastal defences as the primary answer (Royal Commission on Coastal Erosion and Afforestation, 1907–1911). However, in most cases the nature and form of defences were left to individual grass-root approaches due to partisan activity at local level demanding instant built responses: over the last half century the same ethos has been increasingly delivered through more centralized government agencies. These well meaning but engineered-heavy approaches are now viewed as sowing seeds of a disaster to be reaped in the 21st century. It is only in the last decade that the attempts to move towards more proactive consideration of causes of reducing coastal resilience through sustainable protection, rather than meeting only the symptoms of resilience change, has substantially moved the debate on towards methods that may be viewed as resilience enhancement (Figure 11.2, DEFRA, 2001).

The debatable long-term effect of UK engineering protection can be schematically seen in Figure 11.3 (Orford, 1986) in the context of changing coastal cell structure. Human intervention has led to the development of deflected wave sediment cells (source-corridor-sink) linked to longshore sediment disturbances, induced by the sequential introduction of overlapping defence and resource exploitation measures. The loss of foreshore resilience through offshore and longshore sediment leaching associated with engineering structures has been identified on a UK scale by Taylor, Murdock, and Pontee (2004), who through a national GIS consideration of coastal change identify that 61% of the English and Wales engineered coastline show beach rotation or beach steepening. This is defined by the mean low tide position

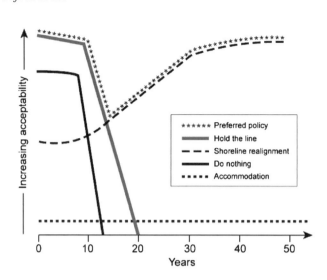

Figure 11.2: The tolerability of each coastal management policy option identified for the UK (DEFRA, 2001), as part of the preferred pathway over the next half century. Note the expected loss of 'hold-the-line' and 'do nothing' options relative to the expansion of 'shoreline re-alignment' (managed retreat) as the dominant option by 2020 AD in the UK. It is recognised that an element of built protection will be retained for the foreseeable future.

spatially retreating landwards (over the last 150 years) at a faster rate than the mean high tide position has altered. This change is explained by the loss of beach face sediment volume, as much as by spatially trying to fix the mean high tide via hold-the-line protection approaches.

An associated protection issue for the British Isles is the large number of estuaries (Davidson et al., 1991) that have experienced rapid reclamation since the 18th century, for both flood defence and the creation of industrial zones based on the break of transport of bulk volume resources. As a consequence of reclamation, estuary tidal prism adjustment has forced offshore distortions in coastwise sediment cell pathways, thereby sequestering shoreline sediments into the near-shore store or further offshore and compounding loss of sediment to down-drift coastal positions (Orford, 1988). A further dimension for the eastern England has been the strategic attempts at closing off small estuaries through dyke building and beach elevation reprofiling to maintain fresh/brackish water habitats (more recently to fulfill obligations to EU directives on birds and biodiversity), which have as a consequence shut down the potential for beach face sediment to enter into contemporary back-barrier and estuary depositional zones.

The recognition of enhancing natural coastal resilience as a contra aim to coastal defence has an old history, but remains a much lower and neglected position within the coastal protection community. Carey and Oliver (1918) recognised attempts to halt the rate of erosion losses by measures to reduce vulnerability by artificially increasing resilience through adding vegetation to dunes and on gravel to entrap mobile sediment. While Strahan (1931) in a debate about coastal morphology, aptly described a situation where 'England is

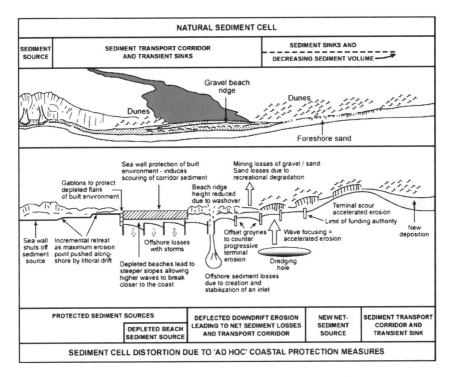

Figure 11.3: A schematic view as to how an original wave-sediment cell of sediment source, transport corridor and sediment sink, is disturbed and deflected by human intervention in pursuit of social and economic development centred upon the coastline. These type of associated coastal defences are symptomatic of much of 19th–20th century urbanised developments along the UK coastline. In most cases sediment deficiencies are experienced at the shoreline over this time period (based on Orford, 1986).

not very large and were it not for its shingle (gravel) beaches, it would be a great deal smaller' while arguing for a more reflective approach to maintaining gravel features. These were isolated voices of reason in the coastal wilderness, given the subsequent continuing issue of UK coastal erosion and flood risk (e.g., 1953 flood event) and the pervasive use of engineered coastal defence as the principal response. Crooks (2004) estimates that 23% of the English coastline is now protected 'soft' cliffs, while sea defences protecting floodplain lowlands account for a further 34% of the coastline length (3740 km).

Natural Controls on Resilience

Resilience changing has natural controls and some of these have been operating on long-term (centennial) scales that appear to be lost under the rash of human intervention, or else problematically enhanced in effect by human intervention. As the rate of relative sea-level change (RSLC) decreases, then it is likely that longshore sediment supply falls. Although the threat of future climate change tends to support ideas of future rising sea levels, it is worth considering the last millennium as a period of relatively low RSLC in much of Britain. In such a period of near-stationary sea level, longshore cells were initially encouraged, but often quickly worked out the limited sediment available, leading to deficit-stressed cells. This sediment-deficit status is often reflected in the switch from drift-aligned to swash-aligned cells and England shows a dominance of the latter with its associated temporary alleviation of loss of resilience through self-organisation. Unfortunately this period of swash-alignment stabilisation coincided with the prolonged spread of Victorian coastal resort development in the 19th century, further stressing the already depleted coastal cells. Given the future expectations for RSLR, then the question of reinvigorated sediment supply to increase resilience can be considered (Orford et al., 2002). However, three further conditions indicate that this renewed supply may not be sufficient:

- Modern coastal protection has been sealing off most of the new longshore terrestrial sediment sources. It will take time for these defences to be reduced sufficiently to improve sediment volume.
- It is likely that the coastal context arising from the centennial lag in past supply will force a future volume lag especially to down-drift positions — such that improving resilience will be spatially dependent.
- Fast RSLR (fast transgressions) can threaten the stability and both the longshore and cross-shore coherence of cells due to the lag in sediment protection.

The last element needs some further consideration as little is known about the way accumulation coasts respond to varying rates of transgression. Future UK sea-level changes when reduced to the scale of accelerating annual rates, are specified as 4–10 mm a^{-1}: between two to six times the current range of UK rates (<2 mm a^{-1}). We recognise that the actual inundation of RSL rise is thought to cause only a small proportion of actual shoreline movement: for example Galvin (1983) provided an early estimate (20%) of eastern US barrier island shoreline change due to inundation alone and that was assuming a RSL rise rate of $c < 2$ mm a^{-1}. This regional figure is now thought to be problematic (G. Stone *pers com*, 2005), but underlies the dominance of other sources of change to shoreline position due to the shifting nature of shoreline morphology, through re-energised and hence re-mobilised sediment pathways and temporary sinks as RSL rises. The real problem is ascertaining how this remobilisation of existing sediment stores will respond to future variable RSL rate changes. In Britain the special element of gravel barriers adds a further problem that Jennings et al. (1998) identified that of rapid breakdown of gravel dominated barriers when RSLR is >8 mm a^{-1}. Thus coastal resilience may undergo catastrophic falls, while associated remobilisation of on/near shore sediment stores and how they will respond to future rate changes to balance such resilience changes is problematic.

The other major element of future climate change for coasts is storminess changes in terms of future number and intensity of cyclones. Though IPCC (Houghton et al., 2001) identifies a wide spectrum of uncertainty of this element at the regional level, Hulme et al. (2002) and Lozano et al. (2004) suggest that the western UK will not necessarily experience a major growth in storm number, but could expect a shift in storm intensity with an increase in larger depressions. Such storms are causes of barrier mobility through overwash generation and barrier retreat. If storminess increases, then the consequent reducing time period between bigger storms can lead to squeeze on upper barrier sediments which are combed down beach/offshore in storms, and have less time (decreasing inter-storm spacing) for fair-weather rebuilding. Storminess also leads to increasing overwashing rates as upper barrier elevation is reduced due to sediment loss. Overall losses of sediment volume from barriers result in smaller barriers that will retreat faster (non-linear) as decadal RSLR increases (Orford et al., 1995), leading to accelerating back-barrier squeeze and possible barrier breakdown.

One of the key elements in UK coastal protection approaches is to try and maintain gravel barriers through artificial reprofiling to maintain crest elevation, and sediment nourishment to build up beach volume. There is concern that neither approach is sustainable. Figure 11.4a

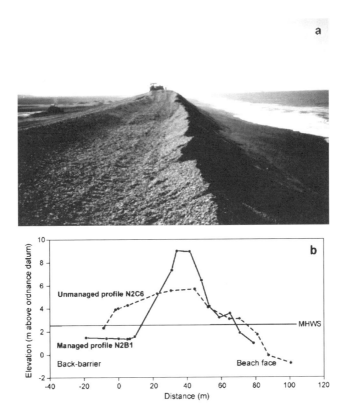

Figure 11.4: (a) Artifical reprofiling of the gravel barrier at Cley (North Norfolk England: photograph courtesy of John Pethick) to act as a flood defence scheme. Such profiling is designed to a specified elevation related to an extreme water level (tide plus storm surge), but it adds to decreasing barrier resilience through forcing a beach profile disequilibrium, as witnessed by the comparison of normal and managed profiles (b). The reprofiling is currently an annual event, indicative of the disequilibrium with respect to profile and storm activity (profiles courtesy of the UK Environment Agency).

shows Cley gravel barrier (Norfolk), being reworked for elevation maintenance to ensure flood protection and enforcing 'spatial' stability. However, this reworking is adding to decreasing barrier resilience through forcing system disequilibrium, as witnessed by the comparison of normal and managed profiles (Figure 11.4b). In particular, extreme storm activity would lower the barrier crest elevation (relative to decreasing volume) and move the barrier landwards by rollover.

Segmentation of Coastal Sediment Pathways

The lesson to be learned from the overuse of coastal protection in the UK is that decreasing coastal resilience will be caused by the consequential deficit in coastal sediment. Illustration of the scale of such sediment deficits for eastern England as a whole can be gauged from Table 11.3, which identifies the relative balance of available coastal sediment and the sediment requirements to meet existing coastal morphology demands engendered through contemporary and future RSL rise. Orford and Pethick (2006) have undertaken a first-order approximation of this sediment budget and identified how with increasing annual RSLR, these sediment demands will slip exponentially into massive deficits by the end of the 21st century.

If these demands are realized then associated loss of coastal resilience and increased PV will ensue. Although increasing PV as a whole is identified there is another dimension to the problem in terms of coastal segmentation. The issue of segmentation is of serious

Table 11.3: Estimated annual sediment availability between Flamborough Head and North Foreland (eastern England) and budget requirement for varying rates of sea-level rise rates.

	Present and future RSLR mm a^{-1}	Cohesive sediment (M)	% Demand of available sediment (+, surplus; −, deficit)	Non-cohesive sediment (M)	% Demand of available sediment (+, surplus; −, deficit)
Sediment availability Mm^3 m a^{-1}		12		1.6	
Sediment demand Mm^3 m a^{-1}	2	3.3	+72.5	1.2	+75
	6	10	+16.6	3.6	−125
	8	13.2	−13.0	4.8	−200
Sea-bed requirement Mm^3 m a^{-1}	2	60	−400		
	6	180	−1400		

Source: After Orford and Pethick (2006).

concern in that a coastal management ethos of the last decade is to consider that coastal zone management should be seen through an integrated perspective. Indeed, the demand to use coastal cells in planning UK shoreline management is central to the concept of integrated coastal development. However the concept of coastal segmentation is based on the perspective that sediment pathways (the integrative dynamic of integrated coastal development) have broken down or are in deficit mode. An accumulation of two centuries of forcing disequilibrium and inducing rotation by coastal engineering and reclamation, means that too much coastal sediment has been sequestered offshore or out of the active littoral system. One might take some comfort that this sediment supply reduction calibrated by the switch from drift to swash alignment has witnessed an increase in coastal resilience through self-organising stability. However this movement is not an end in coastal organisation, as evidence suggests that swash alignment is a precursor to eventual barrier breakdown (Orford et al., 1996) — probably the most serious physical problem for open coast development given the exposed accommodation space opened up to the rear of the barrier.

This problem of reducing resilience is not limited to eastern England. Carter (1988, pp. 208–209) identifies a section of the Ceridigion coast of western Wales with extended coastal cell development. The overall wave power potentials translated into longshore sediment transport loads up to 40k $m^3\ m^{-1}\ a^{-1}$. In comparison, the actual sediment loads realised through the limited sediment availability is less than 1k $m^3\ m^{-1}\ a^{-1}$. Such sediment deficit leads via cannibalisation to cell compartmentalisation, such that wave-sediment cells are increasingly developing at smaller scales (<5 km). Intriguingly, this breakdown of longshore structure might indicate a source of future increasing coastal resilience, as the potential for change under forcing will not show the same scale of lateral communication as would be expected in an integrated system. Although the bulk of the Ceridigion coast shows a natural decreasing resilience, the segmentation of the beaches around Aberystwyth to the north of the Ceridigion section is very much a product of human interference due to coastal defences and harbour entrance piers (Figure 11.1c). Sediment scarcity has led to: a loss of northwards sediment pathway continuity, swash-aligned cells disconnected from the northern coastal drift system, isolated coastal systems being forced by transgressive conditions, and increasing potential of beach and barrier breakdown that has required sediment nourishment as unsuccessful remediation. Barrier degradation is witnessed by the changes in one of the Aberystwyth barrier beaches (Tan-y-bwlch, Figure 11.5a, b), where beach profile (Figure 11.5c) adjustments between 1983–2002 have been observed by Pethick et al. (2003). Figure 11.5d shows the annual rate of migration of specific index contours (6 m, HAT, MHWS and MSL) for profile positions along the barrier (Figure 11.5c). A discernable beach face steepening of this fixed barrier can be seen. This is recognised as 'beach rotation' (Taylor et al., 2004), where in this case, sediment supply to the beach face has been reduced. Figure 11.5e shows the annual rate of mean profile migration, with the inevitable retreat of the barrier ridge in the middle of the bay where sediment is finest, volume is lowest and barrier ridge is as yet undefended. This is inevitably a point where breaching will occur and points to the next spatial threshold to resilience change in the barrier.

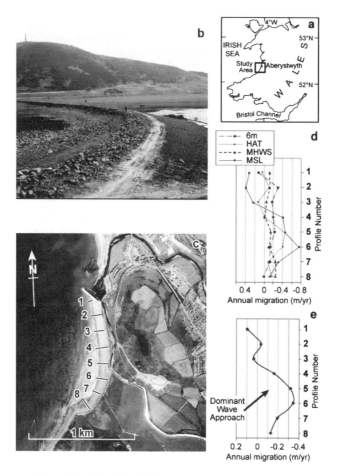

Figure 11.5: Structure (a) and location (b) of a Tan-y-bwlch gravel barrier (Aberystwyth, Ceredigion, west Wales). Barrier degradation is witnessed by the barrier changes between 1983 and 2002 (from Pethick et al. (2003)). (c) identifies profile positions along the barrier from which the annual rate of migration of index contours are shown in (d). A discernable beach face steepening of this fixed barrier can be seen. (e) shows the annual rate of mean profile migration and the indication of diminishing resilience as the swash-alignment deepens to the point of barrier breaching (photographs courtesy of Ceredigion County Council).

Segmentation of the coastal system has decreased the ability for coastal sediment pathways to be re-supplied at the rate required to exceed or match barrier requirements to prevent destabilisation. These barriers are now more likely to migrate at faster rates as their volume decreases, which in turn leads to more rapid barrier breakdown and increasing vulnerability. The segmentation process does however mean that vulnerability changes are likewise spatially constrained to smaller individual coastal segments and have less chance of rippling through the broader coastal system.

How Can Coastal Resilience be Improved by Human Action?

Resilience will rise when sediment is retained in the coastal zone. Future coastal management must work to include a new concept of 'sediment husbandry' as a prime element of practice. This requires two approaches: aim to trap still-available sediment within the coastal zone and reduce *future* accommodation space by encouraging sediment movement onshore and up-estuary to fill potential spaces. How can this be encouraged as a resilience building measure? At this stage the most obvious method is shoreline realignment (Leggette et al., 2004) with an emphasis on trapping sediment otherwise lost to the system. Within the next decade, this mode of response is planned to be the preferred tolerable option in coastal management in the UK (DEFRA, 2001). Its driving ethos is that of sustainability, but this approach would also be the major future resilience building policy (Figure 11.2). However shoreline realignment is essentially undertaken only in estuaries as a function of cohesive sediment supplies. The approach is also open to criticism in that current active realignment schemes are hardly coping with 20th-century induced coastal disequilibria as a function of past protection, let alone for any 21st-century need of multiple/ rolling setbacks to accommodate accelerating RSLC. In 50 years time, a managed retreat that opens up 100 ha of medieval reclaimed marsh on the UK East Coast, would result in a total sediment demand of over 1 Mm^3 at probable rate of 10,000 m^3 a^{-1} on a total resource estimated to be in the region of 12 Mm^3 a^{-1}. This may be a small percentage of the annual supply, but if 10 such managed realignment sites were opened their combined demand would amount to 1% of a resource already outstripped by the demands of the existing coastal habitats. The result must be either that managed realignment sites in 2060 AD would fail to accrete or, depending on location, would divert sediment from existing inter-tidal habitats that would then deteriorate. Thus action is required immediately, while sediment supply is still in positive balance, so as to trap this dwindling resource.

A further cautionary note is that re-alignment in open coast situations, where the emphasis is to open up barriers to cross-shore flows and hence landward sediment movement, has still to be undertaken on any major scale in the UK.

Specific Imperatives for Increased Resilience through Sediment Supply

The current emphasis on fine sediments for re-alignment should not dominate sediment management perspectives. Sand dunes may be regarded as the non-cohesive sediment analogue to salt marsh in that they provide substantial volumes for exchange at times of storm stress on beaches, but the current move to stabilise these areas (especially for recreation purposes e.g. golf courses) should be resisted. Enlightened sediment husbandry would reconnect dunes to their source areas of the inter-tidal beach, removing intervening defences and, more importantly, removing obstacles to renewed deposition, for example bulkheads, boardwalks and vegetation that impede saltation on/off the dunes. Similarly, gravel barrier beaches should be reconnected to their cliff sources areas, removing the upstream groynes that have diverted so much of this sediment resource into offshore sinks.

Source areas themselves must be identified and, where necessary, remobilised. Cliff defences are an obvious contender but more controversial would be removal of cliff vegetation cover, such as plantations of trees, to facilitate slope failures and encourage sediment entry into coastal pathways (Leafe et al., 1990).

The scale of the problem means that we must extend our range of intervention tools far beyond conventional techniques of bypassing existing defences. There is a need to mobilise not only onshore sediment volumes but also to engage with nearshore sources. One of the most critical sediment sinks along the UK coast, but largely ignored by coastal management, is the tidal delta at estuary mouths, which through estuary reclamation and engineering structures have increasingly switched to dominant ebb orientation. These sand bodies trap substantial volumes of coastwise non-cohesive sediment and, in so doing, provide onshore shelter for the movement of cohesive sediment under tidal dominance through an estuary entrance. Tidal deltas require tidal discharges from estuaries as a function of the tidal prism allowed physically into the estuary. In this regard managed realignment may be used to increase estuarine prisms, as well as shifting the time–friction profiles of estuaries to encourage flood asymmetry of tidal flow, with the added possibility of ebb to flood switches in the dominant morphology and hence encourage sediment into the estuary *per se*. More controversial may be the re-opening of former tidal channels that were 'fully reclaimed by artificial barriers', or even the provision of new tidal channels, for example by allowing or even constructing breaches in barrier beaches in attempt to access nearshore sediments.

These approaches to sediment husbandry need to be bedded with planning barrier/beach management at a coastal scale. In the UK that means we should use a behavioural approach to understanding coastal organisation, which places an evolutionary and regional emphasis on beach and barrier development. At the present time, all coastlines in England and Wales are subject to a process of shoreline management planning (SMP; MAFF, 1995). Revisions of these plans are due to be carried by 2008 (SMP Version 2), and are liable to be judged against the behavioural approaches identified by the key government agency (Defra) through their Future Coast programme (Burgess et al., 2002). The imposition of sediment husbandry as a key element in improving coastal resilience and hence decreasing PV has still to be formally accepted in this planning process, but should be seen as a co-supporter of sustainability, which is already at the core of SMP Version 2.

Future Society Scenarios: Controls on Susceptibility and Resilience

Although one can anticipate a general statement as to the value of sediment husbandry in a coastal context, recent work by the UK Foresight programme on the implications of fluvial and coastal flooding and coastal erosion over the next century (Evans et al., 2004) has recognised the differential role of governance on how society may approach future substantial environmental problems, dependent on society's standing on two overlapping continua of governance and values. Figure 11.6 identifies the four generalised domains of these future scenarios. Each scenario can be mapped against IPCC (Houghton et al., 2001) scenarios for climate change. In a broad sense these scenarios can be seen as increasing

Figure 11.6: The potential structure of UK future government — consumerism domain into four scenarios related to scale of climate forcing and hence potential for coastal forcing (Based on Evans et al. (2004) and the constraint basis for the coastal responses identified in Table 11.4).

coastal forcing albeit at differential rates. All of these scenarios therefore show future increasing PV if resilience does not change. It is also evident that there could be varying resilience capacity through human intervention under these future society scenarios. Table 11.4 is an attempt to assess how the management of coastal barriers (as a specific example of coastal morphology dependent on sediment supply) might vary between scenarios. It is observable that different scenarios could lead to different emphasis on human intervention into the coastal system and consequent impacts on barrier vulnerability.

Conclusions: Changing Coastal Physical Vulnerability

Coastal PV is an effect of both natural and human controls on susceptibility and resilience, the balance of which specifies the level of vulnerability. Increasing resilience is the most probably way to effect changes in vulnerability. In the UK, nearly two centuries of human development of coasts supports society's coastal resilience, but has undermined coastal systems resilience by engendering sediment loss from the shoreline due to coastal protection. This basic pattern of engineered-heavy approaches to coastal protection can be identified across many parts of the world's coastline, so that the imperatives for increasing resilience presented in this chapter have wider application than to the UK environment alone.

Decreasing vulnerability can be achieved by supporting moves to coastal self-organisation in the face of sediment reduction at the shoreline. This requires positive intervention within coastal systems towards sediment retention. More probable should be attempts to revitalise

Table 11.4: Variable responses to possible future UK governance/values scenarios, affecting human approaches to coastal vulnerability through protection issues related to coastal barriers. This is one possible example of how future human intervention may affect how society may wish to deal with coastal protection and hence its impact on coastal resilience.

		Values	
		Consumerism	**Community**
Governance	Autonomy	*National enterprise* • Protect barriers with push for agricultural self-sufficiency as economic benefit (as in 1939–1945) • Regional scale activity • Barrier protection for common good	*Local stewardship* • Barriers left untouched under ecological principle and left to retreat • Local self-motivation might pursue barrier stability • Local scale will reduce effective need in integrated coast, but might work in segmented coast • Real estate forcing is minimal
	Interdependent	*World markets* • Maintain barriers for agriculture • No subsidies, so marginal land may allow barrier breakdown as no effort made to support barrier • Real estate values can force barrier protection	*Global sustainability* • Barriers maintained for habitat protection • Variable value on specific habitats relative to international agenda • Not controlled by agricultural return • Real estate carries little weight

Source: Scenarios are from Evans et al. (2004).

sediment pathways by opening up terrestrial sources currently locked out of the coastal pathways by coastal defences. More controversially, we should attempt to capture sediment in onshore sinks, as the major aim for increasing resilience in the face of future susceptibility changes. Without such positive approaches coastal zones are likely to suffer increasing PV in the face of continued protection pressure, let alone by the effects of future climate changes.

References

Adger, W. N. (2000). Social and ecological resilience: Are they related? *Progress in Human Geography, 24*, 347–364.

Adger, W. N., Hughes, T. P., Folke, C., Carpenter, S. R., & Rockstrom, J. (2005). Socio-ecological resilience to coastal disasters. *Science, 309*, 1036–1039.

Burgess, K., Orford, J. D., Dyer, K., Townend, I., & Balson, P. (2002). FUTURECOAST — the integration of knowledge to assess future coastal evolution at a national scale. In: McKee Smith, J. (Ed.), *Proceedings of the 28th international conference on coastal engineering*, American Society of Civil Engineers.

Carey, A. E., & Oliver, F. W. (1918). *Tidal lands: A study in shoreline problems*. London: Blackie and Son Limited.

Carter, R. W. G. (1988). *Coastal environments: An introduction to the physical, ecological and cultural systems of coastlines*. London: Academic Press.

Crooks, S. (2004). The effect of sea-level rise on coastal geomorphology. *Ibis, 146*, 18–20.

Davidson, N. C., d'A Laffoley, D., Doody, J. P., Way, L. S., Gordon, J., Key, R., Drake, C. M., Pienkowski, M. W., Mitchell, R., & Duff, K. L. (1991). *Nature conservation and estuaries in Great Britain*. Peterborough: Nature Conservancy Council.

DEFRA. (2001). *Shoreline management plans: A guide for coastal defence authorities*. London: Department of the Environment, Food and Rural Affairs (DEFRA).

Evans, E. P., Ashley, R. M., Hall, J., Penning-Rowsell, E., Saul, A., Sayers, P., Thorne, C., & Watkinson, A. (2004). *Scientific summary: Volume I — future risks and their drivers*. Foresight; future flooding. London: Office of Science and Technology.

Forbes, D. L., Orford, J. D., Carter, R. W. G., Taylor, R. B., Shaw, J., & Jennings, S. C. (1995). Morphodynamic evolution, self-organisation and instability of coarse-clastic barriers on paraglacial coasts. *Marine Geology, 126*, 63–85.

Galvin, C. J. Jr. (1983). Sea-level rise and shoreline recession. *Coastal zone, 3*, 2684–2705.

Houghton, J. T., Ding, Y., Griggs, D. J., Noguer, M., van der Linden, P. J., Day, X., Maskell, K., & Johnson, C. A. (Eds). (2001). *Climate change 2001: The scientific basis*. Intergovernmental Panel on Climate Change. Cambridge: Cambridge University Press.

Hulme, M., Jenkins, G. J., Lu, X., Turnpenny, J. R., Mitchell, T. D., Jones, R. G., Lowe, J., Murphy, J. M., Hassell, D., Boorman, P., McDonald, R., & Hill, S. (2002). *Climate change scenarios for the United Kingdom*. The UKCIP02 Scientific Report. Norwich: Tyndall Centre for Climate Change Research, University of East Anglia.

Jennings, S. C., Orford, J. D., Canti, M., Devoy, R. J. N., & Straker, V. (1998). The role of relative sea-level rise and changing sediment supply on Holocene gravel barrier development; the example of Porlock, Somerset, UK. *The Holocene, 8*, 165–181.

Klein, R. J. T., & Nicholls, R. J. (1999). Assessment of coastal vulnerability to climate change. *Ambio, 28* (2), 182–187.

Leafe, R., Pethick, J., & Townend, J. (1990). Realising the benefits of shoreline management. *The Geographical Journal, 164* (3), 282–290.

Leggette, D. J., Cooper, N., & Harvey, P. (2004). *Coastal and estuarine managed realignment*. Report C628. Construction Industry Research and Information Association (CIRIA).

Lozano, I., Devoy, R. J. N., May, W., & Andersen, U. (2004). Storminess and vulnerability along the Atlantic coastlines of Europe: Analysis of storm records and of a greenhouse gases induced climate scenario. *Marine Geology, 210*, 205–226.

MAFF. (1995). *Shoreline management plans. A guide for coastal defence authorities*. London: UK Ministry for Agriculture, Fisheries and Food.

Orford, J. D. (1986). Coasts: Environments and landforms. In: P. Fookes, & P. Vaughen (Eds), *Handbook of engineering geomorphology*. London: Chapman and Hall.

Orford, J. D. (1988). Alternative interpretation of man-induced shoreline changes in Rosslare Bay, southeast Ireland. *Transactions Institute British Geographers, 13*, 65–78.

Orford, J. D., Carter, R. W. G., & Jennings, S. C. (1996). Control domains and morphological phases in gravel-dominated coastal barriers. *Journal of Coastal Research, 12*, 589–605.

Orford, J. D., Carter, R. W. G., McKenna, J., & Jennings, S. C. (1995). The relationship between the rate of mesoscale sea-level rise and the retreat rate of swash-aligned gravel-dominated coastal barriers. *Marine Geology, 124*, 177–186.

Orford, J. D., Forbes, D. L., & Jennings, S. C. (2002). Organisational controls, typologies and timescales of paraglacial gravel-dominated coastal systems. *Geomorphology, 48*, 51–85.

Orford, J. D., & Pethick, J. S. (2006). Challenging assumptions of coastal habitat formation in the 21st century. *Earth Surface Processes and Landforms*.

Pethick, J., & Crooks, S. (2000). Development of a coastal vulnerability index: A geomorphological perspective. *Environmental Conservation, 27* (4), 359–367.

Pethick, J. S., Orford, J. D., & Young, R. (2003). *Allt Wen a Traeth, Tan-y-bwlch SSSI and Aberystwyth Frontage Nature Conservation Strategy*. Report for Countryside Council for Wales and Ceredigion County Council.

Royal Commission on Coastal Erosion and Afforestation. (1907–1911). *Report and minutes of evidence 1907; report and minutes of evidence 1909; report 1911*. London: HMSO.

Strahan, H. (1931). Discussion of Lewis, W., the effect of wave incidence on the configuration of a shingle beach. *The Geographical Journal, 78*, 129–148.

Taylor, J. A., Murdock, A. P., & Pontee, N. I. (2004). A macroscale analysis of coastal steepening around the coast of England and Wales. *The Geographical Journal, 170*, 179–188.

Walker, B., Holling, C. S., Carpenter, S. R., & Kinzig, A. (2004). Resilience, adaptability and transformability in social-ecological systems. *Ecology and Society, 9* (2), 5. Available online- http://www.ecologyandsociety.org/vol9/iss2/art5 (last accessed 2 March 2006).

Chapter 12

Promoting Sustainability on Vulnerable Island Coasts: A Case Study Smaller Pacific Islands

Patrick D. Nunn and Nobuo Mimura

Introduction

Covering nearly one-third of the Earth's surface, the Pacific Ocean is largely water yet also contains myriads of islands (Figure 12.1): a few larger ones like Hokkaido (Japan) and New Caledonia, and a host of smaller ones. Most smaller islands are arranged in clusters, many classified as archipelagic (such as Fiji and the Galapagos), others within groups that are more linear (like the Hawaiian Islands and most in French Polynesia). Only a few islands are truly isolated (like Nauru and Niue).

Most islands are located in the southwest quadrant of the Pacific and in lower latitudes, where the processes by which islands originate and endure are most active (Nunn, 1994). Owing to this concentration, most smaller islands have tropical climates and, largely on account of their comparative smallness and remoteness, most such islands are also considered part of the 'developing world'.

Islands are innately more vulnerable to many of the most powerful forces of environmental changes because of their insularity. Islands have larger coastal length to land area ratios than continents. On account of their shapes, many smaller Pacific Islands have some of the highest such ratios of any of the world's landmasses (Table 12.1). In the atoll nations of Kiribati and Tuvalu, for example, every part of the land is coastal, in the sense of both its environmental fabric and its inhabitants being inexorably linked to the sea. On larger islands like some in Fiji and Samoa, most people live within a few kilometres of the shore and most economic activity is located on flat areas along the coast. Threats to coastal environments in the Pacific Islands therefore not only pose threats to most of their inhabitants but also to their means of livelihood, and sometimes their entire island world (Nunn et al., 1999; Pelling & Uitto, 2001).

A significant problem in managing coastal vulnerability on many islands (and in the 'developing world' more generally) is that those charged with management have normally

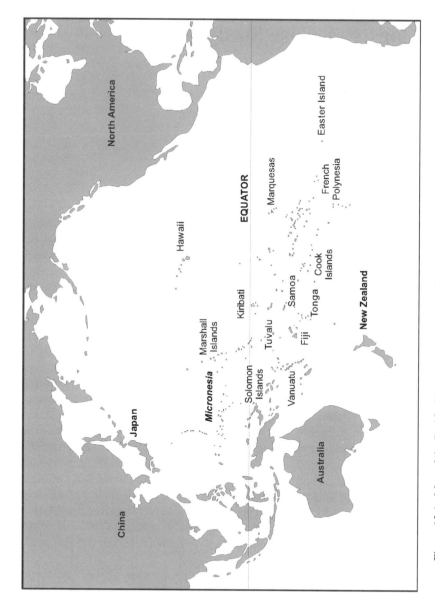

Figure 12.1: Map of the Pacific Island regions showing the principal island groups mentioned in the text.

Table 12.1: Characteristics of selected Pacific Island coasts.

Group/Nation	Land area (km²)	Coastline length (km)	Insularity index (coastline length/land area) × 100	Population density (persons per km²)	GDP per capita (US$)
Cook Islands	240	120	50	88	4998
Federated States of Micronesia	702	6112	871	154	—
Fiji	18,270	1129	6	48	5551
French Polynesia	3660	2525	69	72	4959
Japan	374,744	29,751	8	339	28,699
Kiribati	811	1143	141	122	802
Marshall Islands	181.3	370.4	204	311	2038
Nauru	21	30	143	599	4773
New Caledonia	18,575	2254	12	11	14,231
Niue	260	64	25	8	3543
Northern Marianas Islands	477	1482	311	168	—
Palau	458	1519	332	43	—
Papua New Guinea	452,860	5152	1	12	2051
Philippines	298,170	36,289	12	284	4487
Samoa	2,934	403	14	61	5612
Solomon Islands	27,540	5313	19	18	1571
Taiwan	32,260	1566.3	5	701	17,962
Tokelau	10	101	1010	142	1058
Tonga	718	419	58	151	2182
Tuvalu	26	24	92	435	1079
Vanuatu	12,200	2528	21	16	2823
Wallis and Futuna	274	129	47	57	1907
Means			157	175	5014
Totals	1,245,391	98,424			

Note: Data from CIA World Factbook, December 2003, via in early May 2004, www.nationmaster.com.

been trained in/for continental situations and apply continental-derived solutions to island environments without regard to their considerable differences. Islands are not just miniature continents (Doumenge, 1987) and to regard them as such may result not only in a failure to manage them effectively but also create new unanticipated problems (Nunn, 2004). An early parallel example was the introduction of alien species of plants and animals, particularly within the past 200 years, that have devastated the highly vulnerable native island ecosystems (McNeill, 1999). Today, throughout most tropical Pacific Islands, governments and local communities are protecting their eroding shorelines with 'hard' artificial structures, wrongly believing that these provide a viable long-term inexpensive solution to the problem yet eventually discovering that they are a short-term solution that creates new problems and requires more money to maintain than is available or justifiable (Nunn, 2004).

This chapter first considers the nature of Pacific Island coasts, separating their physical character from human interactions, then goes on to deal with their historical development. A discussion of coastal vulnerability treats traditional (pre-1850) and modern (post-1850) separately, and is followed by a comprehensive account of future changes in the Pacific Islands coastal zone and how this will affect vulnerability and its management. The concluding section focuses on future directions for the management of coastal vulnerability in these islands.

Nature of the Pacific Islands Coastal Zone

To most outsiders, the thought of Pacific Island coasts conjures up an idyllic image of a pristine environment untouched by the harsh 21st-century environmental conflicts that are familiar to people living elsewhere in the world. The reality is quite different. Behind the façade that is sometimes created and often maintained at great cost (typically for tourists), there is a host of coastal-environmental issues in the Pacific Islands that are likely to worsen in the future unless appropriate action is taken.

The Physical Character of the Pacific Islands Coastal Zone

Despite widespread variations, the coasts of most smaller Pacific Islands are: cliffed and commonly of hard rock, for example those found around high limestone islands such as Vava'u in Tonga and volcanic islands like Nuku Hiva in the Marquesas (French Polynesia), or low lying and composed of sediments, commonly unconsolidated, such as are found on atoll islands (*motu*) which comprise most islands in the countries Kiribati, Marshall Islands, Tokelau and Tuvalu, and along the older, more denuded coasts of larger islands, especially around river mouths or adjoining broad coral reefs, as in the Rewa Delta of Viti Levu Island in Fiji.

Cliffed island coasts are generally not vulnerable to the same degree as low-lying coasts, and neither are they generally favoured sites for habitation or economic activity, so they will be discussed here in brief. Some such coasts are indicative of emergence (uplift), particularly high limestone islands located close to convergent plate boundaries; the principal vulnerability of such locations may be to tectonic phenomena, including locally generated tsunami. Similarly, many steeply cliffed volcanic islands in the Pacific are young, and volcanic and related hazards pose some of the greatest threats. There is a tendency for the vulnerability of cliffed islands to be underestimated. Twenty-first-century sea-level change may pose little threat to Niue Island, which is fringed by 23-m high cliffs, yet at least two tropical cyclones (hurricanes or typhoons) in the past two years have overtopped these cliffs causing considerable damage (Figure 12.2).

Low-lying coasts are far more vulnerable to most environmental changes, but variations within this group are important for environmental managers to comprehend. For example, coasts with a high hinterland are inherently less vulnerable than those with none, as on atoll islands. The following components are recognized and discussed separately below: onshore, shoreline, and nearshore. While these are useful for discussion purposes, they form part of an interacting system in which changes in one component frequently affect the others.

Figure 12.2: (a) What remained of the K-Mart (supermarket) on Niue Island after Cyclone Heta in January 2004. Note the huge boulders that were thrown up onto this 23-m platform by the giant waves. (b) The place where Tuapa Church once stood on Niue Island, following the impact of Cyclone Heta in January 2004 (photos by Emani Lui, used with permission).

Onshore Onshore environments of low-lying coasts vary in their vulnerability depending largely on elevation and land use. Extensive coastal lowlands may appear the most vulnerable category, but under present conditions they may offer considerable resilience — an attribute of the natural or human system that reduces its vulnerability to outside forces — in the face of inundation because of the network of drainage channels that may exist there. Thus people continue to occupy the 50-km^2 Navua River Delta in Fiji, despite inundation every five years or so, because the land is fertile and the floodwaters rarely cause lasting damage. Narrow strips of coastal lowland may exhibit heightened vulnerability because threats such as storm surges are focused on comparatively small areas. The villagers of

Navuti, one of the largest on 11-km² Moturiki Island in Fiji, are crowded into a narrow coastal strip and considering re-location should sea level continue rising (Figure 12.3a).

Much also depends on land cover. The existence of a 'buffer zone' of coastal forest is often enough to reduce the vulnerability of any low-lying hinterland (Figure 12.3b). That

Figure 12.3: (a) Navuti Village on Moturiki Island in central Fiji occupies a 40-m broad strip of low-lying coastal flat that lies 10–30 cm above mean high-tide level and is regularly flooded. Behind the village are steep cliffs. (b) The buffer zone of vegetation between the ocean (on the right) and the village of Amuri (left) on Aitutaki Island in the Cook Island. This buffer zone was created, and most of the houses moved father inland than before, following damage by large waves during Tropical Cyclone Sally in 1986 (photos by Patrick Nunn).

said, most such locations in smaller Pacific Islands were targeted early on in the modern era as places for significant settlements (such as Apia in Samoa) or plantation agriculture (such as the sugar cane that occupies most coastal lowlands on the largest islands of Fiji). In such areas vulnerability is higher than it would be were they occupied by only a scatter of subsistence dwellers.

Most such people living on smaller Pacific Islands prefer to occupy narrower coastal strips (see Figure 12.3a) close to productive offshore coral reefs. For this reason, their settlements are frequently away from the mouths of large rivers, which are associated with low ocean-surface salinities detrimental to coral growth. Favoured environments would be an emerged beach and coastal flat as close to the sea as possible, an environmental choice that has continued largely since the first settlers arrived (Nunn, 2005). The dependence of most Pacific Island people on marine foods means that coastal management and mismanagement are long-standing issues on many islands (Nunn, 2003).

Shoreline While some low-lying shorelines are marked by berms or storm beaches, that to some extent shield the hinterland from inundation, many such locations are marked by beaches that slope gently upwards and into a coastal plain at typically around 1.0 m above mean high-tide level. Few beaches on the more densely populated islands are in a 'natural' condition, most showing signs of disequilibrium attributable to both recent environmental changes (such as large waves) or direct human impact (such as beach-sand mining or mangrove clearance, Figure 12.4a–d).

The shorelines of most low-lying coasts in the smaller Pacific Islands are also generally lightly vegetated, in some cases natural but in others a product of human actions, allowing ready people — access to the shore but also sea breezes to aerate the hinterland. Mangrove forest fringing the shoreline was once far more widespread than it is today, much of it having been a casualty of human impact in the past 150 years or so. A 30-m broad fringe of mangrove forest can exert a huge protective influence on the shoreline. It can not only break the force of waves but, being a porous barrier, will also absorb their energy. A study of settlements around the coast of Ovalau Island (Fiji) showed that the two (Bureta and Visoto) where tribal taboos prevented the mangrove fringe being cut down, were experiencing significantly less shoreline erosion than at the front of other coastal settlements on Ovalau where such taboos were never applied (see Figure 12.5, Nunn, 2001a).

Nearshore Most Pacific Islands are tropical and most of their coasts are surrounded by coral reefs. Often reefs fringe the shoreline, sometimes they break the ocean surface a distance offshore as barrier reefs, in which case they are usually separated from the island shoreline by a shallow lagoon. In either case, reefs provide an important zone of physical protection of island coasts from the ocean, something that is particularly important when large waves drive onshore. The force of the 2004 Indian Ocean Tsunami in the Maldives (Indian Ocean) was reduced when it encountered offshore reefs (www.oceansatlas.org).

Yet it is too simplistic to equate the existence of reefs to the effective protection of the coast. For example, much depends on the state of the tide when large waves reach an island coast; reefs may barely offset such waves when the tide is high. It is also pertinent to consider that reefs in a healthy condition that are growing well may provide more effective

Figure 12.4: (a) Shoreline erosion manifested by fallen coconut palms along the back of the beach near Navitilevu Village on Naigani and Island, central Fiji. This coast was formerly fringed by a mangrove forest – the remain of the shore flat on which this grew is exposed in the foreground. The mangroves were cleared enabling waves to reach the back of the beach where they have been causing erosion of the coastal flat (photo by Patrick Nunn). (b) Coastal erosion at Nukui Village, Rewa Delta, Viti Island, Fiji. This coast began experiencing severe erosion in 1995, probably owing to destabilization associated with storm surges (photo by Nobuo Mimura). (c) Coastal erosion of Funafuti Atoll, Tuvalu. This coast has long suffered from serious erosion. People tried to protect the coast by putting stones, concrete blocks, and other materials along it, but with little success (photo by Nobuo Mimura). (d) Beach on Ha'atafu Beach in Tongatapu Island, Tonga. Beachrock develops on the shore face providing protection to the sandy beach (photo by Nobuo Mimura).

protection from large waves than ones that are largely barren, perhaps as a result of coral bleaching. Island coastal managers should realize that the likely increased incidence of coral bleaching over the next few decades (Hoegh-Guldberg, 1999) will reduce the importance of reefs as a barrier to wave erosion. Yet here again the view is more complex. Coral bleaching sometimes results in an increased supply of (reef-derived) sediment to lagoons which will alter sediment dynamics and may, depending on the specific nature of the site, either enhance or reduce the vulnerability of a particular shoreline to wave erosion over a given time period.

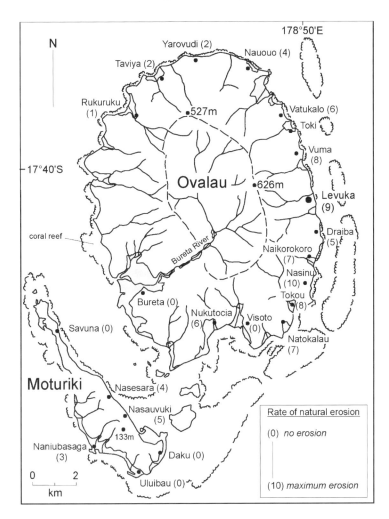

Figure 12.5: Map of Ovalau and Moturiki islands, central Fiji, showing the relative severity of shoreline erosion at every coastal settlement. The severest erosion is along Ovalau's windward (southeast-facing) coasts. Traditional taboos on mangrove clearance have ensured that the mangrove fringe have been preserved at Bureta and Visoto villages, where there is no significant shoreline erosion as a result (after Nunn, 1999a, 2000a).

Just seaward of the shoreline of most low-lying coasts in the smaller Pacific Islands, there is a broad shore flat which is often exposed at low tide, made typically of coral rock covered with sediment. Sometimes this shore flat runs into a fringing reef, of which the reef crest is closest to the ocean surface (low-tide level) and has the most critical protective role. The reef crest and fore-reef slope are the most important areas where calcareous sediment is produced, sediment that is driven landward and builds beaches. Where organic productivity of these reef areas drops, the adjoining beaches may become starved of sediment and begin eroding (Figure 12.6).

Figure 12.6: Shoreline erosion to a tropical Pacific Island resort in July 2003 resulted in the removal of the sand covering the beachrock (darker area on right) and the limestone bedrock (centre). Sand removal to depths of around 60 cm is shown by the exposure of the foundations of the thatched umbrellas. The problem here was traced to non-biodegradable laundry detergent in wastewater being emptied onto the gardens and finding its way to nearshore areas where it produced an algal bloom. The algae smothered the corals, resulting in a lack of sediment production which in turn starved the beach of sand causing the erosion shown. A switch to eco-friendly detergent, the manual clearing of the algae, and the dumping of sand from a nearby cay on this beach appears to have restored the reef-beach dynamics that prevailed in earlier times (photo by Patrick Nunn).

Many low-lying coasts are linked with barrier reefs (rather than fringing reefs) separated from the shoreline by a lagoon, typically reaching depths of 50–100 m. In general terms, the morphology of a barrier reef is similar to that of a fringing reef and is an important source of both lagoonal and shoreline (beach) sediment.

Atolls are rings of coral reef, usually broken in one or more places by a reef pass, that enclose a shallow lagoon, marking the place where a volcanic island subsided. Many atoll reefs (like many barrier reefs) host islands (*motu*) of largely unconsolidated calcareous sand and gravel, formed when large-amplitude (storm) waves drive this sediment up the talus slope below the fore reef onto the broad back reef.

In terms of vulnerability, it is useful to consider the highly variable results of large waves on atoll islands (*motu*). Many such waves cause erosion of *motu*, sometimes obliterating them altogether. Yet sometimes they enlarge *motu* by driving onshore reef detritus that will eventually become accreted along the shores of existing *motu*. Studies of *motu* in Tuvalu showed that most such islands formed from the successive accretion of 'rubble banks' deposited on reef flats during storms (McLean & Hosking, 1991). The causes of the variability may have to do with sediment availability on the reef-island slopes and their form relative to the direction from which large waves approach, but they may also manifest the condition of a particular *motu* shoreline, as suggested for Ontong Java Atoll (Solomon Islands) by Bayliss-Smith (1988).

Human Interactions with Pacific Island Coastal Zones

Ever since Pacific Islands were colonized by humans, they have preferred to occupy the coasts. This is largely because these are often the only places where flat lowlands, suitable for house building and agriculture, exist. Yet this preference also reflects the dependence of most Pacific Island people, even today, on foods that can be hunted or gathered from nearshore and offshore areas. In assessing the vulnerability of Pacific Island coastal zones, it is sometimes easy to overlook the dependence of their occupants on its food resources.

Most people occupying the coasts of smaller Pacific Islands are not integrated fully into their nation's cash economy, and subsistence food production is a critical component of their livelihoods to which most people have no realistic alternative. Pacific Island people are accustomed to interacting with the coastal zone with no detriment to the foods with which it can supply them. This means that the concepts of vulnerability and adaptation, and even of management, often appear alien and inappropriate to community-level decision-makers. In reality of course, Pacific Island people understand vulnerability very well, and have been adapting to changing patterns of resource availability ever since they colonized the islands.

In terms of vulnerability, it often seems that later (possibly more naïve) arrivals on a particular island occupied more vulnerable locations because other (more prized) locations were already occupied. This has been suggested as a reason why 'Polynesians', who settled in the islands of Vanuatu some two millennia after they had been colonized, occupied the smaller, often lower, islands off the larger ones (Nunn et al., 2006). Adaptation often became ritualized in Pacific Island societies through the adoption of totems by particular tribal groups, meaning that there were/are particular food items (commonly types of fish or crustacean) that they would not eat, thereby leaving more of these for a neighbouring group which would, in turn, eschew some food items that the first group could eat. Such practices have long been interpreted as adaptation, following the realization that particular foods were in short supply and/or needed to return to levels that were routinely exploitable (Johannes, 1981; Veitayaki et al., 2003). In modern times, as many such practices have been discontinued, efforts are being made to enforce similar types of adaptation by establishing marine-protected areas in places where unsustainable exploitation of particular environments or particular marine foods has taken place (King & Lambeth, 1999; Veitayaki et al., 2003).

Many foreign-derived coastal management solutions also appear alien and unworkable to many community-level decision-makers in Pacific Islands because of assumptions made about the values of particular components of the coastal environment. Beaches, for example, are highly valued in most 'western' schemes for coastal management, largely for aesthetic and recreational reasons that are often assumed to be universally shared. In most subsistence-based Pacific Island societies, beaches are valued far less, often regarded as no more than rubbish dumps or places onto which boats are hauled (Nunn, 1999a, 1999b).

On most smaller Pacific Islands — those which are not hubs of national development — urbanization and infrastructure are commonly insignificant except where there is an income-generating, site-specific activity such as a large tourist resort or a mine.

Notwithstanding official statements, in both situations the imperative of making money and creating jobs usually takes precedence over issues of coastal environmental impact (Brown, 1974; Turnbull, 2004).

Historical Changes in the Pacific Island Coastal Zone

The coastal zone of Pacific Islands has always been among the most dynamic of the region's environments and, — under the influence in particular of changing climate and sea level, they cannot be regarded as ever having been truly stable (Nunn, 1999a). The principal control on Holocene (the last 10,000 years) coastal development in the Pacific Islands has been sea-level change.

During the Early Holocene (10,000–6000 years ago), sea level was rising and most coasts today exhibit signs of drowning. By inundating the lower parts of valleys that existed on islands prior to the Holocene, the length and form of many island coastlines changed, typically from straight and short to sinuous (embayed) and long. This change created coastal environments that were sheltered and therefore suitable for the establishment of certain ecosystems (such as mangroves), which attracted humans millennia later.

Sea-level rise during the Early Holocene was accompanied by warming of ocean-surface waters that saw the re-establishment and spread of coral reefs through much of the tropical Pacific. In some parts of the region, reefs were able to grow upwards at the same rate as sea level was rising, but elsewhere reef upgrowth lagged behind sea-level rise. This contrast between 'keep-up' and 'catch-up' reefs was important during the Middle Holocene (6000–3000 years ago) when sea level stopped rising and began cutting laterally into island coasts. Those coasts fringed with reefs experienced less erosion than those where reef surfaces lay several metres below the ocean surface (Hopley, 1984; Nunn, 1994).

Towards the end of the Middle Holocene, Pacific sea levels began falling, a process that continued until about 1200 years ago (Dickinson, 2003). It was within this period of time that humans began to colonize smaller Pacific Islands, a coincidence that has been explained by the creation (through sea-level fall) of attractive coastal environments for human settlement (Nunn, 1994; Dickinson, 2003).

There is debate about the degree to which early humans in the Pacific Islands impacted the pristine environments they encountered (Nunn, 2001a). Much of this debate focuses on island interiors, and there is little evidence that early humans had much direct impact on island coasts. Most coastal-environment change during the period 3000–1200 years ago in the tropical Pacific Islands is attributable to sea-level fall and the effects of 'catch-up' reefs finally reaching the ocean surface (Nunn, 2005).

From approximately 1200–700 years BP, sea level began rising as temperatures increased during the Little Climatic Optimum (Nunn, 2000a). This was generally a time when Pacific Island societies, founded largely on a diet of marine foods, became more complex. However subsequently there was a short-lived period of rapid cooling and sea-level fall, named the 'AD 1300 Event' (Nunn, 2000b) that devastated many Pacific Island societies. This was principally due to the effect of sea-level fall on the productivity of reef and lagoon ecosystems. A massive drop in foods available along the coast drove many people inland and offshore (Figure 12.7). During the ensuing Little Ice Age

(600–150 years BP), inland rather than coastal settlement was dominant, and many coasts were affected by increased inputs of terrigenous settlement associated with land clearance for agriculture in upland areas inland. A good example is provided by the Sigatoka Valley on Viti Levu Island in Fiji, where inland hilltop settlements were established following abandonment of coastal settlements during the AD 1300 Event. Land clearance around these hilltop settlements was signalled by charcoal deposition on valley floors, and increased amounts of river sediment load that produced a dunefield along the coast (Kumar et al., 2006).

An upsurge of foreign settlers (mostly European) on many Pacific Islands coincided with the time of warming and sea-level rise (beginning about AD 1800) that has characterized the most recent period of coastal history in the region. It witnessed a gradual re-establishment of coastal settlements and increased exploitation of coastal and marine foods that had formed less important parts of human diets during the Little Ice Age on many

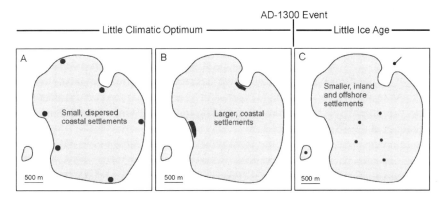

Figure 12.7: Changes in settlement pattern on a typical smaller island in the tropical Pacific during the past 1200 years. More details were given by Nunn (2000c). (a) At the start of the Little Climatic Optimum (Medieval Warm Period) about 1200 years BP (AD 750), most settlements were small and dispersed along island coasts. Owing to low population densities and subsistence lifestyles, most coastal-environmental impacts were low. When the vulnerability of particular areas was exposed, perhaps after an extreme event, most communities would have responded by moving to a nearby, unaffected site. (b) Towards the end of the Little Climatic Optimum about 700 years BP (AD 1250), because of increasing aridity, smaller communities had amalgamated and human impacts on island coasts had become focused on particular areas. (c) During the early part of the Little Ice Age, following the environmental crisis associated with the AD 1300 Event, larger coastal settlements were abandoned in favour of smaller inland or offshore, commonly fortified sites. Many coastal environments were hardly exploited by people for a few hundred years. Land clearance for agriculture in the islands' interiors led to increased fluvial sediment loads and shoreline progradation along island, particular around the mouths of large rivers, and the associated spread of mangrove forests.

islands. The spread of foreign ideas of coastal management throughout the region was fuelled by European colonization of most Pacific Island countries.

Managing Modern Coastal Vulnerability in the Pacific Islands

Modern coastal vulnerability in the smaller Pacific Islands has many facets but also needs to be managed in the context of increasing future vulnerability. The overall challenge to devise management strategies that provide effective solutions to present-day problems of the coastal zone yet also anticipate future changes. For organizational ease, past and present coastal vulnerability is discussed in this section and future changes in the next section.

For most of the time that humans have interacted with the coasts of most smaller Pacific Islands, these interactions have been less sophisticated, less changeable, and altogether less threatening to the natural process regime than many occurring on larger, more densely populated islands in the region today. For this reason, such 'traditional' management practices are discussed separately from 'modern' management practices below.

Traditional Management Practices

Little can be directly known about coastal management during pre-modern times in most parts of the smaller Pacific Islands. Coastal management, if it ever existed, was probably of comparatively low environmental impact, as a result of which no legacy of it has remained in the landscape. Some information has been gleaned from questioning of elderly inhabitants of some modern island coasts and by examining the management practices of people who might still be expected to be using traditional practices (Mimura & Nunn, 1998; Nunn et al., 1999).

In pre-modern times, probably the most widespread response to coastal changes that negatively impacted coastal food resources or coastal habitation, was for the affected community to move either elsewhere on the same island or to another island. This adaptation option was widely available in the past when population densities were far lower and no land was 'owned' in its modern sense. In addition, the investment of coastal communities in building was far less; in fact, the nature of many traditional dwellings suggested that they were also transportable, readily rebuilt elsewhere when necessary. That said, there are examples of places where coastal management in the form of lines of boulders, stick fences, and even the development of buffer zones were created to enhance a site's resilience to coastal erosion (Mimura & Nunn, 1998).

As communities on smaller Pacific Islands became more permanent, as population densities rose, and people began to favour the occupation of particular sites (for defensive reasons or because of cultural issues), so coastal management practices became more visible. Following the AD 1300 Event, the natural defensive qualities of many settlements on smaller offshore islands were enhanced by the construction of stone walls. An example of this practice comes from Lelu Island in Kosrae, Micronesia, where after about AD 1400, people began to use fill to raise the level of the reef flat and to build walled compounds to enhance the site's natural defensive qualities (Athens, 1995). In addition, artificial islands were created across the Pacific at this time. Some islands, like the group of 92 that comprise Nan Madol in Pohnpei, Micronesia, involved foundation and seawall construction of remarkable complexity (Ayres, 1983).

Modern Management Practices

As the modern era began, and with colonial administrators and land 'developers', coastal communities suddenly found themselves on land they 'owned' rather than merely 'occupied' and unable to move freely to adjoining lands or islands because they were now 'owned' by others (France, 1969). This limited the adaptation options available for coastal communities, particularly those threatened by undesirable coastal-environmental changes, and forced them to become more permanent (an administrative convenience for many colonial authorities) and to search for means of sustaining the sites they occupied (Clarke, 1977). This change ushered in the widespread use of artificial shoreline-protection structures, intended to provide enduring protection of particular sites, but also other related management strategies associated with the 'ownership' of a particular site and the surrounding resources.

Many resource-management strategies impinged, and most commonly in negative ways — on shoreline change. Many rural communities, particularly those within range of expanding urban centres, became involved in producing food surpluses that could be sold in nearby markets (Brookfield, Ellis, & Ward, 1985). In many areas this resulted in overexploitation and associated degradation of nearshore resources, as increasingly unsustainable methods of food procurement (such as dynamiting and fish poisoning; Figure 12.8a) were used. Many such nearshore areas are now barren or covered by algae; they produce far less sediment than they once did, thereby depriving nearby shorelines of sand and causing them to erode more rapidly than they might have otherwise.

Some coastal communities which lacked aggregate for road construction or materials for brick manufacture, targeted coastal sources, thereby adding to their vulnerability. Reef rock has been mined for seawall manufacture in many parts of the Pacific Islands, and has also been sold commercially for other purposes (Figure 12.8b). Beach sand mining is a major cause of shoreline erosion on many Pacific Islands. Such practices can generally be monitored, if not controlled, on larger islands and/or those closer to the centres of government in the region (Dollar, 1979). However it is almost impossible to enforce the relevant laws on smaller or 'outer islands' which may be visited by government officials only infrequently and where 'common practice' is followed and only rarely questioned (Biribo & Smith, 1994).

As a result of various factors, the shorelines of most Pacific Islands appear to have experienced net recession over the period that sea level has been directly monitored in the region. This recession, manifested as a landward movement of the shoreline (bounded by mean high-tide and mean low-tide levels), is attributable primarily to sea-level rise of around 10–15 cm within the last 100 years (Wyrtki, 1990; Nunn, 2001b). The site-specific effects of this sea-level rise have varied depending on the character of the particular coast. On many of the lowest-lying, most gently sloping coasts, this sea-level rise has resulted in the inundation of the edges, and in some cases the greater parts, of coastal plains, displacing their inhabitants. Most coastal plains of this kind have also

Figure 12.8: (a) Piles (1 and 2) of *duva* (*Derris elliptica*) roots used for poisoning fish, seen here in a cleared area of mangrove swamps on the island Moturiki (central Fiji) on the occasion of a fish drive intended to gather large quantities of food for Christmas visitors to the nearby village. The use of this poison is illegal because it kills not only the larger (target) fish but also smaller fish and corals (photo by Patrick Nunn). (b) Fragments of reef rock on sale on the side of the main highway, Lami Town, just outside Suva, the capital of Fiji. These rocks have been broken off the fringing reef, causing its physical disintegration and killing associated organisms. The rocks are used primarily for lining pit toilets, for which their permeability is essential. Sale of such reef rock is illegal but commonly practiced (photo by Patrick Nunn).

experienced a landwards movement of the area of saline groundwater (salinization) attributable to sea-level rise.

In the first half of the 20th century, it appears that most coastal communities on smaller Pacific Islands were sufficiently adaptable to be able to absorb the effects of sea-level rise. Since about 1950, as populations and the numbers of permanent buildings and infrastructure, have grown the situation on many such islands has become more problematic. In addition, coastal-environmental problems have been exacerbated by socio-economic problems (such as unemployment) and hinterland demands (such as logging) in many places.

The heightened level of crisis in many places over the last few decades has led to a range of responses, commonly involving seawalls. Most of these have been non-site specific, economically burdensome, inappropriate, and only short-term solutions to a continuing problem (Figure 12.9).

Rather than strategies that are costly and unsustainable, it makes more sense in the 'developing world' to seek those that are cheap and sustainable. The key strategic principal for the smaller tropical Pacific Islands is the (re-)planting of a mangrove forest, ideally about 30 m broad, along the most vulnerable island coasts. Such a mangrove fringe, which may have existed along many such coasts prior to their (recent) human occupation, absorbs the energy associated with large waves, thereby reducing their impact on adjacent shorelines. It also helps stabilize sediments in the coastal zone, often allowing the land to prograde and to become more resilient to erosion (see above). The ecological benefits to coastal-dwelling humans of mangrove forests are immense, since they not only house a diversity of marine-littoral foods, but also have a significant role in nurturing the young of many open-water species.

Other advantages of planting mangrove fringes along vulnerable island coasts are that they are cheap to establish and largely self-renewing, in that physical damage during storms will commonly be repaired by natural re-colonization of affected areas. Such a practice will also apply to areas of mangrove forest cleared by people, typically for firewood.

Despite these benefits of mangrove replanting, there is still considerable resistance to it as a coastal management tool in the Pacific Islands (Lal, 2003). The main problem is that it is time consuming; it may take 25 years for a 30-m broad mangrove fringe to reach maturity. For many governments, which have keen for re-election every 4–5 years, seawalls are a better short-term measure even though their long-term effectiveness is generally poor. For this reason (among others), most mangrove replanting schemes in the Pacific Islands have been driven by partnerships between local communities and non-government organizations, sometimes informed by regional organizations like SPREP (South Pacific Regional Environment Programme) and USP (the University of the South Pacific) (Tutangata & Power, 2002).

The Management of Extreme Events

Many of the problems discussed above have been exacerbated locally by the impacts of extreme events: most commonly tropical cyclones (also known as hurricanes or typhoons) but also including large waves (tsunami and storm surge), earthquakes, and volcanic eruptions. For reasons of space and generality, only large-wave impact is discussed here.

Figure 12.9: (a) Detail of a wall on Thulusdhoo island, Maldives, to show the ways in which coral rock is utilized in such construction when there are no other sources of hard rock available. A similar situation obtains throughout the smaller Pacific Islands. (b) Remains of the seawall at Yadua Village, Viti Levu Island, Fiji. This seawall was built to counter shoreline erosion but collapsed and was renewed several times before being abandoned. Note how the wall has been broken and how waves have thus got behind it and caused erosion of the coastal platform on which the village is built. This village is now replanting mangroves along part of its shoreline with the help of Japan-based non-government organization OISCA (photos by Patrick Nunn).

Historically, every few years, most island coasts in the western tropical Pacific, and increasingly as a result of sea-surface warming those in the central tropical Pacific, experience uncommonly large waves (perhaps 10 m amplitude) during the summer cyclone season. Many such waves wash a considerable distance inland, and the strongest have destructive impacts on coastal buildings and infrastructure, including artificial shoreline-protection structures. In countries like Japan, many such structures are built not only to repel waves at high tide but also giant waves generated during tropical storms or by seafloor earthquakes. As for most of the countries affected by the 2004 Indian Ocean Tsunami, this degree of shoreline protection, however desirable, is simply not affordable for most Pacific Island nations. They must absorb the impacts of larger-than-normal waves on a regular basis without appropriate engineering solutions.

The ways in which coastal dwellers on most Pacific Islands have coped with the effects of such waves varies, depending largely on both the proximity to assistance-delivery nodes (such as ports and airstrips) and the priority attached by aid donors and central government to particular affected communities. In practice, this usually means that the remoter communities on smaller islands have to depend more on their own resources and resourcefulness, while those communities closer to national centres tend to await outside help.

In December 2002, 5-km^2 Tikopia Island — one of the remotest in the eastern Solomon Islands — was hit by Tropical Cyclone Zone and 'all the coastal villages were washed away'. The inhabitants sheltered in caves during the storm, which brought 10-m waves onshore, but soon afterwards set about rebuilding, not expecting immediate outside aid. But two more tropical cyclones — Gina and Ivy — hit Tikopia within the next 14 months and the inhabitants' self-sufficiency is weakening (www.afap.org).

Tropical Cyclone Heta reached hurricane strength on 1 January 2004 before running across the north coast of Upolu and Savaii islands in Samoa (Croad, 2004). The seawalls constructed by local communities were generally broken up while designed 'riprap' systems (funded by national or international funds to protect the circum-island roads) proved more resilient.

Evaluation of the Effectiveness of Modern Coastal-Zone Management in the Smaller Pacific Islands

As in many more peripheral areas of much of the 'developing world', coastal-zone management in the smaller Pacific Islands is often assumed to fall under the umbrella of national-government (top–down) legislation. However in practice it is usually community driven, sometimes with the assistance of non-governmental organizations.

The disparity illustrates one of the greatest challenges facing environmental managers in the smaller Pacific Islands and, more generally, in the 'developing world'. Aid donors and national governments, following models of coastal-zone management that operate successfully in the 'developed world', assume that legislation and top–down approaches in general will produce the desired response at the local level. Much aid given to Pacific Island nations for coastal management accepts this basic model uncritically.

In practice, the situation is quite different. Most Pacific Island governments employ insufficient, appropriately trained people to enforce and explain the relevant legislation and as a result it is often almost uniformly ignored. Yet since the development and enactment of environment legislation is a popular conduit for aid money to reach cash-strapped Pacific Island (and other) governments, great efforts are made to maintain it.

What happens in reality in the smaller Pacific Islands is that local communities recognize a problem affecting their coasts and, usually without any reference to government officials or legislation, devise and implement their own solutions. To be effective, recommended coastal-management solutions in the smaller Pacific Islands need to acknowledge that different pathways of decision-making prevail in such places (Nunn, 2004; Turnbull, 2004).

Likely Future Changes in the Pacific Island Coastal Zone

Physical Changes

The Pacific Islands will be greatly affected by climate change and sea-level rise induced by global warming during this century. As the degree of global warming varies with the emission of greenhouse gases (GHGs) such as CO_2, it eventually depends on how human society and its economic activities develop in the future.

Global warming induces climate change such as changes in precipitation patterns, and intensity and frequency of cyclones. There have been greater numbers of droughts and heavy rains in the past decades linked to the El Niño–Southern Oscillation (ENSO). Though the interrelationship between global warming and ENSO is not yet clear, if the present trend towards increased intensity of El Niño continues in this century, more frequent droughts and heavy rains may occur in the Pacific Islands.

A major concern for the Pacific Islands is accelerated sea-level rise. Global mean sea level is estimated to rise between 9 and 88 cm by 2100 according to the IPCC third assessment report. As the past rate of sea-level rise is about 2 cm per decade in the Pacific, the estimated maximum increase is some four times larger.

Another concern is changes in tropical cyclones: specifically, their frequency, intensity, seasonality, and geographical range. It is suggested that, higher sea-surface temperature accelerates evaporation of sea water, providing more energy to generate stronger cyclones (Emanuel, 1987). Though there is not yet convincing evidence in the observed record of changes in tropical cyclone behaviour, Walsh (2004) summarizes the current understanding of the effects of global warming on tropical cyclones. Regarding the maximum tropical cyclone intensities, it will increase 5–10% by around 2050 and this would be accompanied by an increase in mean tropical cyclone intensities. As a result, peak precipitation rates are estimated also to increase by 25%. Though there will be no significant changes in regions of formation, formation rate will change in some regions. This is strongly influenced by ENSO as well.

Recent calculations presented a more concrete estimate, using a 20-km mesh, high-resolution global atmospheric model (Oo021_ et al., in review). The tropical cyclone frequency observed in the warm-climate experiment is reduced by about 30% globally, compared to the present-day-climate experiment. However, tropical cyclones in the warmer world tend to be more long lasting with the decreased frequency of short-lived cyclones (less than seven days), and the number of strong tropical cyclones increases. The maximum surface wind speed of the stronger tropical cyclones generally increases by 8.9 m/s in the Northern Hemisphere, and by 5.4 m/s in the Southern Hemisphere. These suggest a possibility of higher risks of more persistent and devastating tropical cyclones in the Pacific in a future warmer climate (Emanuel, 2005).

The multiple actions of these changes in climate and sea level will certainly bring about serious effects on the natural environment and human society in the Pacific Islands.

Human-Linked Changes

The societies of the Pacific Islands are also changing. The previous sections have shown that societies have long developed, and maintained, unique lifestyles adapted to the natural environment of the islands. In the traditional way of life, social structure was strongly based on community support networks and a subsistence economy was predominant. Societal changes have increased since a western culture was introduced during the period of colonialization. Population growth, the importance of the cash economy, migration of people from outer islands to urban centres (periphery to core), growth of major cities, and development of modern industries such as tourism, have been changing the traditional lifestyles in the Pacific Islands.

These changes have affected the resilience and vulnerability of the society. Mangrove clearance for better access to the sea, made coasts more susceptible to high waves and other external forces from the sea. Ironically, land reclamation and seawall construction might weaken the protection of local communities, though they were measures intended to accommodate increased population. In the cities, migration from outer islands in addition to the natural population growth has meant that some people have to live in low-lying, environmentally vulnerable areas. These are all examples of increased vulnerability caused by human-linked changes. Such changes in lifestyle and gradual disintegration of traditional communities will weaken traditional human support networks (based on extended family interactions), which has been a major component of social resilience in the Pacific Islands.

Recent economic globalization will accelerate such tendencies. It may negatively impact the economic position of developing countries in the Pacific on the whole, as they are economically weak and remote from the world economic centres. During this century, the trend of globalization will continue, and it will have many effects on Pacific Island peoples.

Effects of Accelerated Sea-Level Rise and Climate Change

In this century, climate change and sea-level rise will proceed at the same time as Pacific Island societies change, and there is no purpose in discussing future scenarios for one while ignoring the other. Sea-level rise will expand the inundated areas of coastal flats, and cause more frequent flooding by storm surges and tsunamis. Some villages want to move to inland or higher places under such a situation, or, if people want to continue living on the same places, needs for seawall construction will increase to protect their villages. In cases where there are insufficient land available behind their villages, for relocation friction will occur with other villages, which may cause social instability. In atoll islands that are extremely low, a major portion of the islands may be inundated or frequently flooded in the future. Funafuti, the capital of Tuvalu, experiences serious inundation due to seepage of sea water through coral rock during the periods of high tides and, while not unprecedented during the history of the human occupation of Pacific Islands, is anathemic to most of their modern inhabitants. The frequency of inundation has increased recently. If such a situation is exacerbated by accelerated sea-level rise, the habitability of atoll islands will be seriously limited, and their inhabitants may need to move to other islands or even other countries. This is the ultimate impact of expected sea-level rise and climate change (Barnett & Adger, 2003).

Sea-level rise will also cause further intrusion of sea water into fresh water aquifers. There is a high possibility that salt-water intrusion will be superimposed on droughts as a result of El Niño, so that future management of water resources and agriculture faces very serious challenges. Higher sea-water temperatures will cause more frequent coral bleaching, which together with sediment deposition on coral reefs, will reduce their productivity and biodiversity.

In the future, coastal zones will face increased land-based pressures, such as land reclamation for living, tourism development, construction of ports and coastal roads, driven by population growth and economic development. This means that during this century, the coastal environment will experience more pressures leading to degradation from both land and sea in the Pacific Islands. These impacts have serious implications for coastal zone management.

Challenges for Effective Coastal-Zone Management

Based on the understanding of the present situation and future changes mentioned above, coastal zone management in the Pacific Islands has two major challenges; one refers to technical solutions for effective coastal protection and the other involves developing management strategies relevant to smaller Pacific Islands.

The distinctive feature of the Pacific Islands is that they have an integrated system of natural coastal protection (Figure 12.10). As waves — the major external force on the

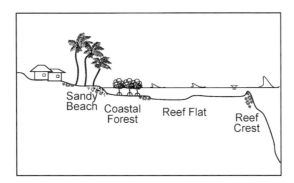

Figure 12.10: Integrated system of natural coastal protection in the Pacific Islands. Waves break first at the reef crest, and continue to lose energy as they propagate across coral flats and mangroves. Sandy beaches are often armoured by beachrock.

islands' coasts — approach the shore, they break at the reef crest, which reduces their energy. Then the energy continues to be reduced as the waves propagate across coral flats and mangroves. When they finally reach the shoreline, they often encounter sandy beaches armoured by hard beachrock. This natural system is the key to maintaining and protecting the coastal environment especially sandy beaches, in spite of the fact that the rate of beach sediment supply is generally slow as it depends on the erosion and therefore ultimately the growth of biogenic structures such as coral reefs and algal ridges.

It is important that the natural barriers spread across this wide area reduce wave energy in this integrated spatial dimension. The failure of the past human practices has been to protect coasts at a given point on the present shoreline by constructing vertical seawalls, rather than trying to enhance natural protection across the entire offshore area.

The most appropriate way of coastal protection in the Pacific Islands is to enhance such natural protective functions as much as possible. In particular, the existence of wide reefs and mangrove fringes are essential (Vanualailai & Mimura, 2004). However, mangroves need a few decades to mature after planting, and there are places where mangrove planting is not an appropriate option to solve the today's problem of coastal erosion. In such a case, artificial structures may need to be introduced. Knowledge of modern coastal engineering is still effective if it is applied under careful consideration of the specific nature of the coastal environment of the Pacific Islands. For example, even if a vertical seawall has to be constructed, protection of its foot by rubble or concrete blocks is very important to prevent wave reflection and hence erosion in front of the seawall. When a groyne or jetty is planned, care should be made not to impede the longshore sediment transport, otherwise serious erosion in the downdrift side will take place. However, there is no complete set of engineering knowledge that is peculiarly relevant to the Pacific Islands. Its development is an area for the future (Barnett, 2001).

It is important to understand that natural systems such as coral reefs and mangroves cannot prevent the effects of long-period waves such as storm surges and tsunamis. Neither are they effective in preventing inundation from long-term sea-level rise. Countermeasures against such changes are limited to the construction of dykes surrounding villages or relocation of them to safer lands.

Engineering solutions should be placed in a wider context of integrated coastal zone management (ICZM). In the face of climate change and sea-level rise, the challenges to ICZM include;

- disaster prevention,
- recovery from disasters,
- water management for drinking and agriculture,
- management of coastal resources, and
- protection of coastal ecosystems.

It is important to note that it is impossible to prevent all the disasters including the unprecedented strong cyclones and extraordinarily high waves. In such cases, indirect measures (other than direct coastal protection are needed), including observation and early warning of tropical cyclones, evacuation to safer areas, and recovery after the damages of cyclones and storm surges. For such actions, the mutual support system in a community plays a major role. As already mentioned above, societies in the Pacific Islands have a long tradition of maintaining social support networks, which is the major element of their resilience to environmental changes. Therefore, another challenge is how to develop ICZM in ways that can enhance these traditional values and incorporate indigenous knowledge (Barnett, 2001).

Conclusions: Future Directions for Coastal-Zone Management in the Pacific Islands

Recent scientific research has increasingly revealed that climate change and sea-level rise pose significant threats to the global environment, particular in those places where humans are most numerous. If the changes proceed as estimated, it is likely that unprecedented environmental changes and devastating natural disasters will happen. In the Pacific Islands, sea-level rise threatens the safety and foundation of living in coastal villages, particularly in areas such as water resources and food supply. In addition, the possibility of increased tropical-cyclone intensity is a further threat. Economic planning for future development and important infrastructure such as seaports, air ports, coastal roads, and urban facilities are also exposed to their effects. The question of how to respond to such threats is a major challenge for governments, local communities, and even ordinary people in the region.

At the same time, the Pacific Islands have a long history of successfully adapting to various environmental changes. This history shows that people in the region recognized and utilized the resilience of island environments. Coasts of the islands are surrounded mostly by coral reefs and mangrove forests, and this natural system has provided highly effective protection to the coasts, which is a source of resilience. Therefore, the most serious implication of climate change and sea-level rise is that they damage these systems thereby reducing the natural resilience on smaller Pacific Islands. On the society side, resilience emerges through strong community ties that provide mutual support to individuals and families, particularly when a community faces a crisis like the serious damage following a tropical cyclone. Communities have also been a basis for the successful management of coastal resources and environments.

Such history shows that the societies of the Pacific Islands were not passive ones, subject only to the effects of environmental changes including climate variability, but rather they adapted to changes in an active manner. Lessons learnt from such history will give examples for future planning of adaptation to climate change and sea-level rise in this century.

ICZM has been long identified as a response measure to sea-level rise. However, the concept of ICZM has been developed in the western 'developed' world, and focuses on role of policies of the central and local governments and modern technology. However, it seems more effective to raise community awareness than to develop policy in the Pacific Islands, an approach that acknowledges the nature of island-archipelagic countries and their societies (Barnett, 2001). There is at present in Pacific Island countries little awareness of the validity of this approach and its effectiveness in addressing the impacts of climate change and sea-level rise. An ICZM relevant to the Pacific Islands should:

- focus on ways to increase natural and social resilience of islands,
- be community-based, and
- combine the traditional management and modern scientific and engineering knowledge to protect coasts.

The new ICZM should also ensure wise use of coastal resources so that it can contribute not only to respond to climate change and sea-level rise, but also to sustainable development of the islands. When such an ICZM is developed, the Pacific Islands will have a tailor-made solution to the threats of climate change and sea-level rise.

References

Athens, J. S. (1995). *Landscape archaeology: Prehistoric settlement, subsistence, and environment of Kosrae, eastern Caroline Islands, Micronesia*. Honolulu: International Archaeological Research Institute.

Ayres, W. S. (1983). Archaeology at Nan Madol, Ponape. *Bulletin of the Indo-Pacific Prehistory Association, 4*, 135–142.

Barnett, J. (2001). Adapting to climate change in Pacific Island countries: The problem of uncertainty. *World Development, 29*(6), 977–993.

Barnett, J., & Adger, W. N. (2003). Climate dangers and atoll countries. *Climatic Change, 61*, 321–337.

Bayliss Smith, T. (1988). The role of hurricanes in the development of reef islands, Ontong Java atoll, Solomon Islands. *The Geographical Journal, 154*, 377–391.

Biribo, N., & Smith, R. (1994). *Sand and gravel usage, South Tarawa, Kiribati, 1989–1993*. Preliminary Report 75, South Pacific Applied Geoscience Commission (SOPAC).

Brookfield, H. C., Ellis, F., & Ward, R. G. (1985). *Land, cane and coconuts: Papers on the rural economy of Fiji*. Canberra: Department of Human Geography, Research School of Pacific Studies, Australian National University.

Brown, M. J. F. (1974). A development consequence: Disposal of mining waste on Bougainville, Papua New Guinea. *Geoforum, 18*, 19–27.

Clarke, W. C. (1977). The structure of permanence: The relevance of self-subsistence communities for world eco-system management. In: T. Bayliss-Smith & R. Feachem (Eds), *Subsistence and survival*. London: Academic Press.

Croad, R. (2004). Samoa's coastline in the aftermath of Cyclone Heta. *Coastal News — Newsletter of the New Zealand Coastal Society, 27*, 1–2.

Dickinson, W. R. (2003). Impact of mid-Holocene hydro-isostatic highstand in regional sea level on habitability of islands in Pacific Oceania. *Journal of Coastal Research, 19*, 489–502.

Dollar, S. J. (1979). *Sand mining in Hawaii: Research, restrictions, and choices for the future*. Sea Grant Technical Paper, UNIHI-SEAGRANT-TP-79-01. Honolulu: The University of Hawaii Sea Grant Program.

Doumenge, F. (1987). Quelques contraintes du milieu insulaire. In: F. Doumenge & M. F. Perrin (Eds), *Iles Tropicales: Insularité, "Insularisme"*. Bordeaux: CRET, Université de Bordeaux III.

Emanuel, K. A. (1987). The dependence of hurricane intensity. *Nature, 329*, 483–485.

Emanuel, K. A. (2005). Increasing destructiveness of tropical cyclones over the past years. *Nature, 30*(436), 686–688.

France, P. (1969). *The charter of the land.* Melbourne: Oxford University Press.
Hoegh Guldberg, O. (1999). Coral bleaching, climate change and the future of the world's coral reefs. *Review of Marine and Freshwater Research, 50,* 839–866.
Hopley, D. (1984). The Holocene "high energy window" on the central Great Barrier Reef. In: B. G. Thom (Ed.), *Coastal geomorphology in Australia.* London: Academic Press.
Johannes, R. E. (1981). *Words of the lagoon fishing and marine lore in the Palau District of Micronesia.* Berkeley, CA: University of California Press.
King, M. G., & Lambeth, L. (1999). *Fisheries management by communities.* A manual on promoting the management of subsistence fisheries by Pacific Island communities, Noumea, New Caledonia: Secretariat of the Pacific Community.
Kumar, R., Nunn, P. D., Field, J. E., & de Biran, A. (2006). Human responses to climate change around AD1300: a case study of the Sigatoka Valley, Viti Levu Island, Fiji. *Quaternary International, 151,* 133–143.
Lal, P. (2003). Economic valuation of mangroves and decision-making in the Pacific. *Ocean and Coastal Management, 46,* 823–844.
McLean, R. F., & Hosking, P. L. (1991). Geomorphology of reef islands and atoll motu in Tuvalu. *South Pacific Journal of Natural Science, 11,* 167–189.
McNeill, J. R. (1999). Islands in the Rim: Ecology and history in and around the Pacific, 1521–1996. In: D. O. Flynn, L. Frost, and A. J. H. Latham (Eds), *Pacific centuries: Pacific and Pacific Rim history since the sixteenth century.* London: Routledge.
Mimura, N., & Nunn, P. D. (1998). Trends of beach erosion and shoreline protection in rural Fiji. *Journal of Coastal Research, 14,* 37–46.
Nunn, P. D. (1994). *Oceanic islands.* Oxford: Blackwell.
Nunn, P. D. (1999a). *Environmental change in the Pacific basin: Chronologies, causes, consequences.* London: Wiley.
Nunn, P. D. (1999b). Pacific island beaches — a diminishing resource? *Asia Pacific Network for Global Change Research, 5*(2), 1–3.
Nunn, P. D. (2000a). Illuminating sea level fall around AD 1220–1510 (730–440 cal yr BP) in the Pacific Islands: Implications for environmental change and cultural transformation. *New Zealand Geographer, 56,* 46–54.
Nunn, P. D. (2000b). Environmental catastrophe in the Pacific Islands about AD 1300. *Geoarchaeology, 15,* 715–740.
Nunn, P. D. (2001a). Ecological crises or marginal disruptions: The effects of the first humans on Pacific Islands. *New Zealand Geographer, 57,* 11–20.
Nunn, P. D. (2001b). Sea level change in the Pacific. In: J. Noye, & M. Grzechnik (Eds), *Sea-level changes and their effects.* Singapore: World Scientific Publishing.
Nunn, P. D. (2003). Nature-society interactions in the Pacific Islands. *Geografiska Annaler, B 85,* 219–229.
Nunn, P. D. (2004). Through a mist on the ocean: Human understanding of island environments. *Tijdschrift voor Economische en Sociale Geografie, 95,* 311–325.
Nunn, P. D. (2005). Reconstructing tropical paleoshorelines using archaeological data: Examples from the Fiji Archipelago, southwest Pacific. *Journal of Coastal Research, 42*(Special issue), 15–25.
Nunn, P. D., Baniala, M., Harrison, M., & Geraghty, P. (2006). Vanished islands in Vanuatu: New research and a preliminary geohazard assessment. *Journal of the Royal Society of New Zealand, 36,* 37–50.
Nunn, P. D., Veitayaki, J., Ram Bidesi, V., & Vunisea, A. (1999). Coastal issues for oceanic islands: Implications for human futures. *Natural Resources Forum, 23,* 195–207.
Oouchi, K. J., Yoshimura, J., Yoshimura, H., Mizuta, R., Kusunoki, S., & Noda, A. (2006). Tropical cyclone climatology in a global-warming climate as simulated in a 20 km-mesh global atmospheric

model: Frequency and wind intensity analysis. *Journal of the Meterological Society of Japan, 84,* 259–276.

Pelling, M., & Uitto, J. I. (2001). Small island developing states: Natural disaster vulnerability and global change. *Environmental Hazards, 3,* 49–62.

Turnbull, J. (2004). Explaining complexities of environmental management in developing countries: Lessons from the Fiji Islands. *The Geographical Journal, 170,* 64–77.

Tutangata, T., & Power, M. (2002). The regional scale of ocean governance: Regional cooperation in the Pacific Islands. *Ocean and Coastal Management, 45,* 873–884.

Vanualailai, P., & Mimura, N. (2004). Present situation of coastal protection system in island countries in the South Pacific. *Proceedings of the 2nd international conference — Asia and Pacific Coasts 2003,* 29 February–4 March 2004, Makuhari, Japan, World Scientific.

Veitayaki, J., Tawake, A., Aalbersberg, W., Rupeni, E., & Tabunakawai, K. (2003). Mainstreaming resource conservation: The Fiji locally managed marine area network and national policy development. In: H. Jaireth, & D. Smyth (Eds), *Innovative governance, indigenous people, local communities and protected areas*. New Delhi: Ane Books (International Union for Conservation of Nature and Natural Resources).

Walsh, K. (2004). Tropical cyclones and climate change: Unresolved issues. *Climate Research, 27,* 77–83.

Wyrtki, K. (1990). Sea level rise: The facts and the future. *Pacific Science, 44,* 1–16.

www.afap.org/apcedi/archive/2004_02_01_archive.html (last accessed 8th March 2005).

www.oceansatlas.org/servlet/CDSServlet?status=ND03MTY4NyY2PWVuJjMzPSomMzc9a29z (last accessed 8th March 2005).

Chapter 13

Managing Coastal Vulnerability and Climate Change: A National to Global Perspective

Robert J. Nicholls, Richard J.T. Klein and Richard S.J. Tol

Introduction

Coastal societies are widely seen as 'vulnerable' to sea-level rise and climate change, often without any serious analysis of the implications of such a rise, especially the potential to adapt. The term vulnerability and vulnerability assessment has also been commonly applied to coastal systems in the context of climate change (McFadden, this volume), most probably reflecting this concern.

This paper explores this 'coastal vulnerability' to climate change and its links to climate and coastal policy. The main focus are the impacts of sea-level rise which has been most investigated to date, although it is recognised that other aspects of climate change could have profound effects on coastal vulnerability, most particularly more intense coastal storms. The chapter examines vulnerability at broad scales meaning national, regional and global scales, which contrast with some of the other more detailed case studies considered in this book. The results are pertinent to the societal response to human-induced climate change, including exploring the need for action, and deciding what policies are most appropriate. In broad terms, the policy choices are fairly simple, comprising combinations of two possible actions:

1. Reducing greenhouse gas emissions, which is conventionally termed mitigation of climate change, linking to climate policy on emissions which is an international issue actively being debated within the United Nations framework convention on climate change (UNFCCC) (http://unfccc.int/2860.php), and resulting instruments such as the Kyoto Protocol.
2. Changing human behaviour in the face of climate change, which is conventionally termed adaptation again linking to climate policy on adaptation under the UNFCCC, and also national and sub-national coastal policy, which sets the context under which adaptation will occur.

As climate change is impacting an evolving world, the chapter first considers the changing coastal zone, including current trends for coastal areas such as growing population and urbanisation, as well as the potential for climate change and sea-level rise. It then explores the notion of vulnerability and scale and the meaning of vulnerability at regional and global scales. This includes the relevant metrics and parameters with which to quantify and explore vulnerability. Using this perspective, the impacts and responses to sea-level rise are considered based on two recent EU projects: (1) DINAS-COAST (building on the paper by Hinkel and Klein, this volume) and (2) Atlantis (Tol et al., accepted). These address sea-level rise through the 21st century, and a new assessment of extreme sea-level rise (>1 m/century) using an existing model, respectively. Collectively, these new studies provide an improved perspective on the vulnerability of coastal areas to sea-level rise, including the limits to both human adaptation and mitigation in terms of managing coastal vulnerability.

The Coastal Zone in the 21st Century and Beyond

The coastal zone has changed profoundly through the 20th century, primarily due to growing populations and economies with a strong urbanising trend, and these changes are continuing at a rapid pace. In 1990, the near-coastal population[1] was 1.2 billion people, or 23% of the world's population living at three times the global-mean density (Small & Nicholls, 2003). Human settlements are also preferentially located close to the world's shoreline, including most larger cities, which means that the world's economy is also concentrated in the coastal zone (Turner et al., 1996; Nordhaus, in press). With continuing trends of population growth and urbanisation, this suggests the development of large lengths of urbanised coastal fringe as existing coastal centres merge over the 21st century, especially in the developing world. The Boston–Washington and Osaka–Tokyo axes, as well as the Mediterranean coasts of Spain, France and Italy are existing examples. Given these profound changes, both in the coastal zone, and within the associated catchments, the coastal system is experiencing multiple stresses, and sea-level rise and climate change are just one element of increasing coastal vulnerability (e.g. Crossland et al., 2005; McFadden, this volume).

The Special Report on Emission Scenarios (SRES) (Nakicenovic & Swart, 2000) was developed to create climate change scenarios, but also provides a way of characterising those worlds so that consistent vulnerability assessments can be made (e.g. Arnell et al., 2004). The SRES report develops a range of possible future world states based on diverging storylines about the future. Hence it develops quantitative estimates of the socio-economic drivers of greenhouse and aerosol emissions, including factors such as population, GDP and technology, and in turn emission scenarios. This provides a consistent input to both climate models and impact assessment models. Under the SRES scenarios, if coastward migration continues at present rates, the near-coastal population could rise to between 2.4 and 5.2 billion people by the 2080s (Nicholls & Lowe, 2004). This is double to more than four times the present levels, and would amount to a third or more of the global population. Economic change may be even more profound and under the SRES scenarios the global economy grows 9–21 times by the 2080s, with GDP/capita rising 3–14 times, and income inequality diminishing in all cases. Of course changes that differ from the SRES scenarios are also possible, but not widely analysed.

---

[1] The near-coastal zone is the area with 100 km horizontally and 100 m vertically of the shoreline.

In addition to the above changes, human-induced climate change is expected to cause profound changes around the world through the 21st century and beyond. This will have important consequences for most coastal residents, even if there is a widespread protection/ armouring response. Due to the concentration of people in the coastal zone, together with important ecosystems, changes in the coastal zone are an important consideration of climate change, and the effects of sea-level rise add a dimension of change not felt away from the coast.

During the 21st century, global-mean sea-level rise will likely be less than 1 m (Church et al., 2001), although given concentrations of people and infrastructure in coastal areas, this still has a high-impact potential (McLean et al., 2001). Threatened areas include coastal cities (Jacobs et al., this volume), deltaic lowlands (Sanchez-Arcilla et al., this volume), small islands (Nunn & Mimura, this volume), and coastal ecosystems (Woodroffe, this volume). If we expand the timescale beyond the 21st century, a larger sea-level rise is possible under some emission pathways due to the destruction of the Greenland and west Antarctic ice sheets, among other changes (e.g. Vaughan & Spouge, 2002; Oppenheimer & Alley, 2004; Lowe et al., 2006). Both large rises in sea level (over long timescales) and rapid rates of rise of sea level (over shorter timescales) may be important in triggering impacts and responses that would not be evident with smaller rises.

Vulnerability Assessment and Scale

The notion of vulnerability is often defined and investigated at the level of the individual or household (e.g. Hinkel & Klein, this volume), and there is a premise that bottom-up 'placed-based' studies are the best way to approach impact and vulnerability analysis (Schröter et al., 2005). While this approach yields important insights at these scales, the notion of vulnerability also has important meaning at more aggregated scales. The emergence of a range of global datasets combined with GIS software for their management is facilitating such analyses (e.g. Vafeidis et al., 2004), providing important insights for a range of policy purposes. While the main focus in this chapter is climate policy and related discussions within the UNFCCC, there are many other applications such as the global World Bank hazard 'hotspot' study which identifies locations subject to multiple hazards (Dilley et al., 2005), or the structured approach to flood analysis used in the UK foresight analysis of flood and coastal defence (Evans et al., 2004a, 2004b). For climate change issues, assessments include global assessments of the benefits of climate mitigation for climate policy purposes (e.g. Tol, 2004), more detailed national and sub-national assessments to promote better coastal management (Nicholls et al., in press), comparative analyses of adaptive capacity (Yohe & Tol, 2002), or to identify 'hotspots' of vulnerability where action is most needed (Hinkel & Klein, this volume). In this paper we take this more aggregated view of vulnerability, and consider the concerns of the UNFCCC process. This means that we must be able to consider both strategic mitigation and adaptation decisions over many decades or even centuries up to global scales.

When considering vulnerability measures at collective versus individual scales, we have an analogy with climate and weather. While there is much commonality in the terms and approaches as the same overall phenomena is being described, in detail the aggregated view of climate requires a different perspective to the instantaneous view of weather. Hence, we might recognise vulnerable countries, regions or cities, but broad-scale analysis does not provide any insight on vulnerability below the scale of analysis (even though differential vulnerability to varying degrees is apparent in all human societies and natural systems). It is noteworthy that this issue of scale and vulnerability assessment has received little systematic attention in climate change impacts research — something that needs to be addressed in the future.

The appropriate metrics and parameters for vulnerability analysis will vary with both scale of analysis and the questions being posed (Nicholls et al., in press). They include impact metrics such as inventories of the number of people threatened, or the area of land threatened; or economic metrics such as the direct economic value of land threatened, adaptation costs, with increasing interest in full economic analyses which include direct and indirect effects of sea-level rise (Darwin & Tol, 2001; Bosello et al., 2004). While a national analysis may count the number of properties lost to sea-level rise, an individual household would be interested in the probability of its property being lost, as well as loss of transports links and other amenity, and how that would affect the mortgage, distances to school, work and shops, etc. Relative crude descriptions of behaviour would suffice for an aggregate analysis — such as the Bruun rule for erosion or the perfect market for its economic effects — but at the local level, more detailed, complicated and data-intensive models are required (even if, when aggregated again, the results are similar). Similarly, a local adaptation analysis will consider the methods and costs in great detail, as a specific project is being designed, while a broad-scale analysis will seek to stylise adaptation in simpler terms such as 'hold the line' (with dikes) versus 'retreat'.

Uncertainty in broad-scale vulnerability assessment also needs to be considered across scales. At broad scales, while the availability of data is improving significantly (e.g. Vafeidis et al., 2004), the uncertainties in many global datasets remain significant. Socio-economic data are generally more poorly developed than physical data, and for example gridded distributions as opposed to national-scale distributions of economic output are only just beginning to become available (see Nordhaus, in press). The description of impact processes is also generalised and this introduces further uncertainties. Lastly, at broad scales, validation/verification of the assessment methods presents significant challenges as suitable datasets are limited or even unavailable. While it has to be an iterative process, validation/verification across broad-scale models for vulnerability assessment needs to receive more systematic attention.

Thus the questions that can be addressed in broad-scale assessments concern broad impacts and responses rather than details of implementation. Rather than focus on absolute quantities, it is better to consider the order of magnitude of the results and the relative magnitude of change to the modelled baseline. Further, as many of the errors are random, the global estimates will have the highest precision as these estimates will benefit most from aggregation. As one considers smaller areas so the uncertainties concerning impact metrics will progressively rise: this restates the above statement about the inherent resolution limits of these types of methods.

Analysing Impacts and Responses to Sea-Level Rise

The impacts and responses to sea-level rise have been considered extensively in earlier reviews (e.g. Bijlsma et al., 1996; McLean et al., 2001; Nicholls, 2002). Table 13.1 summarises the main physical impacts of sea-level rise. These in turn lead to a range of

Table 13.1: Major impacts and potential adaptation responses to sea-level rise.

Physical impacts of sea-level rise	Example adaptation responses (P - Protection; A - Accommodation; R - Retreat)

Inundation, flood and storm damage
    Surge (sea)
    Backwater effect (river)
	Dikes/surge barriers (P)
	Building codes/floodwise buildings (A)
	Land use planning/hazard delineation (A/R)

Wetland loss (and change)	Land use planning (A/R) Managed realignment/forbid hard defences (R) Nourishment/sediment management (P)
Erosion (direct and indirect change)	Coast defences (P) Nourishment (P) Building setbacks (R)

Saltwater intrusion
    Surfacewaters
    Groundwater
	Saltwater intrusion barriers (P)
	Freshwater injection (P)
	Change water abstraction (A)

Rising water tables and impeded drainage	Upgrade drainage systems (P) Polders (P) Change land use (A) Land use planning/hazard delineation (A/R)

For a discussion on protection, accommodation and retreat see Klein et al. (2001) and Nicholls and Klein (2005).

direct and indirect socio-economic impacts such as loss of land and buildings, or loss of tourist amenity or increasing flood risk. Analysis based on economic optimum methods suggests a widespread protection response to such impacts (Nicholls & Tol, 2006), although as shown in Table 13.1, a much wider range of adaptation measures are available. Hence, Table 13.1 illustrates the wide scope of sea-level rise impacts and responses and hence of any vulnerability assessment of sea-level rise. In practise, existing studies have focussed on a sub-set of these natural system effects, and the treatment of adaptation has been limited, or it has simply been ignored. Hence, existing vulnerability assessments of sea-level rise are nearly all incomplete.

In this chapter, we look at some of these issues via new results on coastal vulnerability using the new DIVA[2] model to explore vulnerability to sea-level rise over the coming century, and the FUND[3] model to explore impacts and adaptation under extreme sea-level rise scenarios exceeding 1 m/century over timescales (>100 years) — impacts that might realised by collapse of the West Antarctic Ice Sheet (Tol et al., accepted). Both of these models are based on algorithms designed for aggregate analysis of vulnerability, as defined in the previous section.

---

[2] DIVA (*dynamic interactive vulnerability analysis*) model developed within the DINAS-COAST Project (DINAS-COAST Consortium, 2004).
[3] The model used is version 2.8 (national resolution) of the FUND (*climate framework for uncertainty, negotiation and distribution*) model, an integrated assessment model of a wide range of climate change impacts (Tol, 2004). Here we only analyse the impacts of sea-level rise.

DIVA considers most of the impacts of sea-level rise, together with three selected adaptation approaches where a generalised broad-scale approach is feasible (Figure 13.1; Table 13.2). It is the most comprehensive tool available for looking at sea-level rise impacts, and combines a new global database on coastal zones with a suite of algorithms authored by a group of experts within the DINAS-COAST research consortium (Hinkel & Klein, this volume). Unlike earlier models which only resolved broad regions or at best-individual countries as the base polygons of analysis (Hoozemans et al., 1993; Tol, 2004), DIVA has divided the world's coast into 12,148 segments of variable length, and developed a database containing about 100 parameters based on this functional typology (Vafeidis et al., 2004; McFadden et al., in press). A major benefit of DIVA is that consistent perspectives can be developed across the range of sea-level rise impacts, including a range of protection responses, from no protection to total protection. An intermediate 'optimal' protection option from an economic benefit–cost perspective is included.

FUND considers a smaller set of impacts comprising total (dry) land loss (summing the effects of erosion and inundation) and wetland loss assuming optimal protection from an economic perspective (Figure 13.2). The underlying data is derived from the global vulnerability analysis (Hoozemans et al., 1993) with some updates, while the impact algorithms have been developed first by Fankhauser and then Tol. The FUND model is well understood and lends itself to the sensitivity analyses that were required to explore extreme sea-level rise scenarios over >100 year time spans (Nicholls et al., 2005).

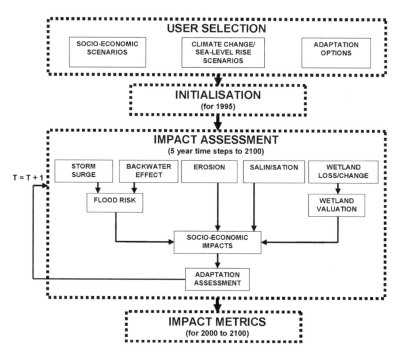

Figure 13.1: Schematic view of the operation of the DIVA tool (see Hinkel & Klein, this volume for more details and an explicit definitions of the constituent models).

Table 13.2: The four major physical impacts of sea-level rise, plus the adaptation approaches that are considered in the DIVA tool.

Physical impacts of sea-level rise	**Adaptation approach**
Inundation, flood and storm damage	
Surge (sea)	Hard protection (via dikes)
Backwater effect (river)	
Wetland loss (and change)	Sediment nourishment
Erosion (direct and indirect change)	Beach nourishment
Saltwater intrusion	
Surfacewaters	Adaptation not considered
Groundwater	

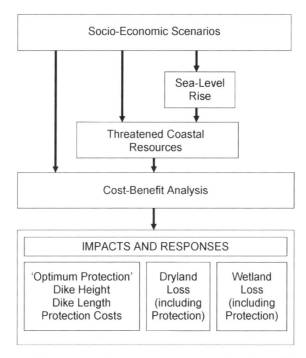

Figure 13.2: Summary of the sea-level rise impact assessment methodology within the FUND model.

In both cases, and like all other broad-scale models, the model outputs are impact metrics which describe factors such as land area loss, number of people affected and economic costs of impacts or adaptation. The vulnerability analysis is conducted externally to the model using expert interpretation via tools such as vulnerability profiles as recommended in the original Common Methodology (IPCC CZMS, 1992).

Impacts and Vulnerability Analysis

Some sample results developed with the new DIVA tool are presented to illustrate these broad-scale tools. The examples used here concern coastal flooding so they are only illustrative of the DIVA tool. A number of papers on aspects of the DIVA tool are in preparation which will provide more technical details. No mitigation scenarios are considered here, but the DIVA tool can easily be linked with mitigation scenarios to analyse what impacts are avoided and hence explore the benefits in terms of reduced vulnerability.

Coastal Flooding at the Start of the 21st Century

Coastal flooding due to storm surges has been examined extensively since the first global vulnerability assessment (GVA) by Hoozemans et al. (1993). DIVA estimates that the coastal flood plain[4] population in the year 2000 was 220 million people, which is similar to the GVA estimate of 200 million people in 1990. Taking estimates of flood defence standards, it was estimated using this data by Nicholls et al. (1999) that 10 million people/year would experience flooding: validation against national scale assessments supported this estimate. DIVA estimates about 3 million people/year experience flooding in 2000 which is still broadly consistent with the earlier estimates.

Unlike the older methods, DIVA allows these results to be examined at a number of scales from global to sub-national administrative units. Figure 13.3(a) shows the regionalised coastal flood plain population, which is concentrated in Europe and parts of Asia. Figure 13.3(b) shows the regionalised number of people flooded/year. They are concentrated in China, south and south-east Asia: Europe disappears due to the current high flood defence standards which means that the risk of flooding is relatively low, despite high exposure. Figure 13.3(c) shows the same data as Figure 13.3(b) by nations — four nations contain 73% of the people flooded by storm surge each year —Bangladesh (12%), China (14%), India (11%) and Vietnam (35%). Figure 13.3(d) looks at this region at the level of the administrative unit. At the administrative level, seven provinces experience flooding of >100,000 people/year due to surge, representing 37% of the global total: Barisal, Chittagong and Dhaka (Bangladesh), West Bengal (India), Jiangsu (China), and Ho Chi Minh and Ben Tre (Vietnam).

When compared with our understanding of the distribution of current coastal flooding, these results look credible. While the people who are being flooded will be well aware of the fact, DIVA provides a global perspective across the problem that did not previously exist, and helps to objectively identify those regions that are presently experiencing significant flooding (i.e. hotspots for coastal vulnerability).

Coastal Flooding at the end of the 21st Century

The DIVA sea-level rise scenarios are given in Table 13.3. For each SRES emissions pathway, a low, medium and high scenario of global-mean sea-level rise was estimated reflecting

---

[4] In all the analyses, the coastal flood plain is defined as the area below the 1 in 1000 year flood elevation.

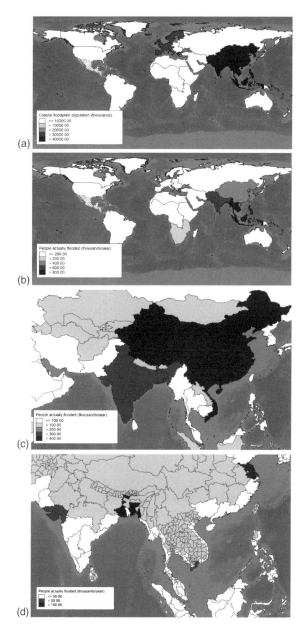

Figure 13.3: Sample results from DIVA: coastal flooding in 2000. (a) Flood plain population; (b) incidence of flooding at a regional scale; (c) incidence of flooding at a national scale and (d) incidence of flooding at a sub-national scale.

Table 13.3: Global-mean sea-level rise scenarios for 1990–2100 for each SRES scenario.

SRES socio-economic scenario	Sea-level rise scenario (m)		
	High	Medium	Low
A1B	0.89	0.44	0.21
A1FI	1.07	0.52	0.25
A1T	0.75	0.36	0.17
A2	0.96	0.47	0.22
B1	0.72	0.35	0.16
B2	0.74	0.36	0.17

Note: High, medium and low refer to climate sensitivity to a given emissions scenario.

estimates based on low to high climate sensitivity: the rise from 1990–2100 ranges from 0.16 to 1.07 m for the full set of scenarios.

Tables 13.4 and 13.5 show the exposure and the experience of flooding in 2100 for these scenarios assuming two quite distinct adaptation options for the full range of scenarios within DIVA 1.0

- *Constant protection* in which the economically optimum dike height in 1995 remains constant with time, to illustrate the consequences of no protection upgrade.
- *Optimum protection* based on benefit–cost analysis to select an optimum dike height (and hence costs).

Table 13.4 shows that the main control on the population living in the coastal flood plain are the socio-economic scenarios — numbers vary much more between the three socio-economic scenarios (A1/B1, A2 and B2) than between the sea-level rise scenarios. The largest population occurs under the A2 scenario with the largest global population. The lowest population occurs under the B1 scenario which has the smallest global population and the numbers are not substantially higher than the 220 million people estimated for 2000. If coastward migration continues, as is currently observed, the exposed population would be higher than shown here (see Nicholls, 2004).

Table 13.5 shows the number of people who actually experience flooding taking account of the exposed population and the defence standards. Assuming constant protection, flooding becomes more and more frequent and by 2100, it is estimated that a significant part of the exposed population is being flooded every year: the number of people flooded increases up to 79 times compared to the 2000 baseline. Hence, assuming no defence upgrade the impacts are quite catastrophic, most especially for high scenarios of sea-level rise. The A2 world experiences the largest impacts, even though sea-level rise is higher in the A1FI world (Table 13.3). However, if we assume optimum protection, despite sea-level rise the number of people estimated to be flooded is reduced compared to 1995 in all cases. This reflects the significant economic growth in all the SRES futures, and people becoming more risk adverse, so the standard of defences that people choose improves even though sea levels are rising. Note that the A2 world still has the highest impacts, illustrating the general point that vulnerability is not a simple function of sea-level rise.

Table 13.4: Global population of the coastal flood plain (millions) in 2100. The results are independent of assumptions about adaptation.

SRES socio-economic scenario	Coastal flood plain population (millions)		
	Sea-level rise scenario		
	High	Medium	Low
A1B	271	252	244
A1FI	279	256	244
A1T	265	248	240
A2	588	542	518
B1	264	248	239
B2	378	355	343

Note: The population scenario for B1, A1B, A1FI and A1T is the same and any differences in the table reflect differences in the sea-level rise scenario.

Table 13.5: Global estimates of people flooded in 2100 (millions/year) assuming constant protection and economically optimum protection, respectively.

SRES socio-economic scenario	People flooded (millions/year)		
	Sea-level rise scenario		
	High	Medium	Low
A1B	202/0.6	113/0.5	35/0.5
A1FI	225/0.6	149/0.5	48/0.5
A1T	186/0.5	85/0.5	27/0.5
A2	473/1.9	284/1.3	98/1.2
B1	183/0.6	84/0.5	24/0.5
B2	283/0.8	136/0.7	44/0.7

Table 13.6 shows the estimated dike investment costs under the two protection scenarios. By definition they are zero under constant protection, while under optimum protection, substantial annual defence investment is required, increasing with both the sea-level rise scenario and the magnitude of exposure. Maintenance costs for dikes are not considered in DIVA.

In general, taking the example of coastal flooding, DIVA shows that potential impacts of sea-level rise are enormous and that they generally increase with the sea-level rise scenario. However, it also shows that they are sensitive to the development pathway and the A2 world appears to experience greater impacts under both the protection options considered here, even though the A1FI world experiences the largest rise in sea level. This reflects that of the SRES scenarios, the A2 world has the largest population (so the biggest exposure), and the smallest GDP/capita (so the lowest ability to adapt), agreeing with several

Table 13.6: Global estimates of sea dike investment costs in 2100 (billions of US dollars/year) assuming constant protection and optimum protection options respectively.

SRES socio-economic scenario	People flooded (millions/year)		
	Sea-level rise scenario		
	High	Medium	Low
A1B	0/84	0/45	0/27
A1FI	0/123	0/58	0/31
A1T	0/61	0/32	0/20
A2	0/148	0/89	0/66
B1	0/56	0/37	0/14
B2	0/70	0/37	0/19

previous analyses (Nicholls, 2004; Nicholls & Tol, 2006). Thus, the analysis confirms that coastal vulnerability is a product of the wider development context of a coastal society or world, as well as the magnitude of sea-level rise, which may not be a surprising result. However, protection (and by implication adaptation in general) could greatly reduce the number of people who experience flooding. If we believe that society makes economically rational decisions, the analysis suggests that under the SRES worlds, the occurrence of coastal flooding will diminish as people become more wealthy (and risk adverse). The general conclusion is that coastal societies will probably have more choice in their response to sea-level rise than is widely expected.

Impacts Under Extreme Sea-Level Rise Scenarios (>1-m rise)

For rises up to a 1-m rise, impacts and responses can be considered as linear functions of sea-level rise, which is a broad assumption underpinning the DIVA tool. For larger rises in sea level, certain impacts become increasingly important, especially land loss and the non-linear aspects of both impacts and adaptation need to be considered if studies are to be meaningful (Nicholls et al., 2005). For instance, people and infrastructure are not distributed uniformly with elevation above sea level, as most impact models assume, and protection costs rise as the square of the defence height, due to geometric considerations, while in areas subject to depth-limited waves costs rise further as nearshore breaking is reduced by rising sea level (Townend & Burgess, 2004).

A rise in sea level exceeding 1 m is increasingly likely as we look further into the future (Nicholls & Lowe, 2006). As an adaptation experiment, the Atlantis project examined the consequences of a hypothetical rapid rise in sea level due to the destruction of the West Antarctic Ice Shelf (see Vaughan & Spouge, 2002). In the Atlantis Project, a range of timescales of destruction were considered, but with most focus being on a total collapse (causing a 5-m global rise in sea level) over 100 years, with an aim to examine societal response to an extreme sea-level rise event (Tol et al., accepted). The likelihood of such an event is not known with any precision, and the 5-m/century rise may even be impossible — however, in terms of understanding impacts and adaptation this analysis has yielded important insights and raised questions which would not have emerged from more traditional analysis. The study comprised case studies of the Thames Estuary (London), the Netherlands and the Rhone delta (France), as well as a global study using FUND. The FUND model was tuned to real population and GDP distributions with elevation using the best-available global datasets, and non-linearities in protection costs were considered via a bilinear step in costings, including a detailed sensitivity analysis (Nicholls et al., 2005).

The global FUND analysis suggests large areas of the world's coasts will be abandoned, but this is not a universal response. Continued protection of some of the world's developed coast is predicted and about one-third of the world's developed coast (or a length of about 250,000 km) is estimated to be protected even for the most extreme scenario (over 5-m/century, when the net sea-level rise is considered), and unit defence costs two orders of magnitude higher than today. This suggests that it is economically rational to protect London and the Netherlands, conflicting with the case studies, which had a more ambiguous response, with retreat being possible, maybe in a rather chaotic and forced manner. Hence, one scale of analysis is missing something important. The case studies could be underestimating the potential and resolve to adapt to long-term change. Equally, FUND (and by implication DIVA) may not capture all the key processes that drive decisions concerning protection versus retreat. Economically optimum decisions are a useful benchmark, but the real world usually lacks the perfect information implied in the economically optimum decisions, and additional criteria will influence decisions. Factors identified in the case studies such as indecision, lack of consensus and a loss of confidence for coastal residents and investors could trigger a cycle of increasing risk, declining investment and blight, and hence trigger quite different responses to an economically optimum analysis, based on present starting conditions and assuming rising living standards.

Barnett and Adger (2003) have discussed socio-ecological thresholds in determining human response to sea-level rise on small islands, and historical analogues from islands in the Chesapeake Bay support their model (Gibbons & Nicholls, 2006). The Atlantis work suggests that there may be important thresholds in coastal vulnerability to sea-level rise, and these are related to indirect as well as direct effects of sea-level rise. This is an area for further research, but needs to be considered when making interpretations concerning coastal vulnerability.

Managing Broad-Scale Vulnerability

The vulnerability assessment methods that we have discussed here are tools for policy steerage — in the context of the UNFCCC process, the goal is to look into the future to see how the world might evolve under realistic scenarios of sea-level rise and other change, and hence understand the potential impacts in the coastal zone. Further we can explore strategic options such as adaptation and mitigation, and hence determine in broad terms how different policies and combinations of policies might reduce these impacts. Coastal vulnerability is the product of both this impact potential and the potential of policies to manage the impacts. These approaches also show that coastal vulnerability is to some degree a product of human choice and there is significant potential to manage this vulnerability.

The impact models that we have used show that there is a large and growing exposure in the world's coastal zones to a range of hazards such as flooding during storms. Sea-level rise will increase the impact potential significantly and in terms of direct impacts, they are overwhelmingly negative: but in economic terms there will be winners and losers, with larger countries generally gaining an advantage (e.g. Darwin & Tol, 2001; Bosello et al., 2004). However, if optimum protection measures are assumed impacts are greatly reduced. In this optimist case, the costs of sea-level rise are protection costs in populated areas, with the physical impacts of sea-level rise being realised in areas with limited population such as the Arctic regions of the USA, Canada and Russia. However, pessimists might point to a number of areas of concern about how the future might unfold (see also Nicholls & Tol, 2006):

- All the socio-economic scenarios considered here assume significant economic growth across the world, and reduced inequality of wealth between countries. Lower growth and/or greater regional differences in development, such as the continued failure of development for Africa, would almost certainly lead to more vulnerable situations than suggested here, even if greenhouse gas emissions and hence sea-level rise was consequently reduced by the lower development.
- Protection is not automatic and presently much of the world is ignoring sea-level rise in coastal management. This suggests that increased protection may be a reactive process to real rather than forecast events (e.g. New Orleans and Hurricane Katrina). This view of the future suggests a higher incidence of coastal impacts and disasters through the 21st century, than if the proactive approach to response that is implied in the benefit–cost analysis was adopted.
- Protection costs against sea-level rise may have been underestimated, especially for deltas and small islands, which are widely seen as the most vulnerable settings to sea-level rise (McLean et al., 2001; Nurse et al., 2001). This is compounded by the potential for more extreme events, which were not evaluated here.
- Important impacts such as those on water resources have not been considered here — at the least this will raise response costs, and in the extreme could be a trigger for retreat. Quantitative exploration of this area at regional and global scales is urgently required.
- Rapid sea-level rise (>1 m/century) is unlikely but still plausible, and the resulting impacts may overwhelm the capacity to protect and encourage a widespread coastal blight and decline.
- Lastly, is protection maladaptation which will increase impacts in the longer term (Smit et al., 2001): while protection benefits the existing population immediately, it will often promote development in vulnerable areas where flood depths and the consequences of a flood event are rising with sea level. This is a difficult question which will depend critically on the timeframe of the analysis, including attitudes to issues such as discounting and intergenerational equity (Arrow et al., 1996) and the magnitude of sea-level rise.

Interestingly, the national foresight study found that portfolios of response options that mixed combinations of retreat and protection appeared the most effective strategy for managing growing flood risk in the UK (Evans et al., 2004a). This was based on multi-criteria analysis and an explicit consideration of sustainability and preserving choice for future generations. However, while Europe is grappling with ideas such as managed realignment, and hybrid responses look likely (e.g. Nicholls & Klein, 2005) in much of Asia, coastal protection and even advance in accreting areas is still the norm (e.g. King & Adeel, 2002; Li, Fan, Deng, & Korotaev, 2004), suggesting that different areas of the world may choose quite different adaptation responses to sea-level rise, especially over the next few decades.

While this discussion has focussed on adaptation, mitigation of climate change can also manage coastal vulnerability. However, it slows but does not stop sea-level rise as historic emissions will contribute to rise for many centuries. The ultimate commitment to sea-level rise (over many centuries) could be significant, depending on total greenhouse gas emissions and the uncertain ice sheet response to the resulting warming. The full-policy implications of this constraint are only recently being understood by coastal policymakers and managers. It means for coastal zones, mitigation alone will not manage the rising risks of sea-level rise, and there is in effect an ongoing 'commitment to adaptation' if coastal vulnerability is not to progressively rise. (Nicholls & Lowe, 2004, 2006).

In conclusion, while adaptation has a great potential to reduce impacts, large or rapid rises in sea level appear to present significant challenges for adaptation. Mitigation reduces the risks of such changes, especially over many decades and longer. Thus, there are several timescales to managing coastal vulnerability: adaptation has immediate benefits in reducing vulnerability, while mitigation reduces vulnerability in the long term.

In this chapter, we have focussed on sea-level rise, and this work can be much further developed and improved. The general principles discussed here could be applied to other coastal issues, including the multiple stresses that characterise many coastal management problems from local to regional scales. A move to analysis of global change impacts (which includes climate change and other human-induced impacts and their interaction) is an obvious extension of the DIVA concept that would improve our understanding and management of coastal vulnerability across this range of scales (cf. McFadden, this volume).

Conclusions

Sea-level rise and the coastal implications of climate change are worthy of concern as they have the potential to cause significant coastal impacts, and other coastal stresses are often increasing the adverse consequences of sea-level rise and climate change. The policy response to sea-level rise cannot be based on either climate mitigation or coastal adaptation alone: to manage coastal vulnerability through the 21st century and beyond will also require a long-term commitment to adaptation combined with mitigation to reduce the future rate and ultimate rise in sea level.

Integrated vulnerability assessment, as presented here, is essential to analyse this problem as the responses require policy action at broad scales: nationally, regionally and globally. Mitigation is clearly a global-scale activity, but adaptation requirements also need ranking and focussing of the limited available aid. Small islands stand out in all the analyses as highly vulnerable with limited capacity to respond, but other areas such as deltas may also require assistance for adaptation to sea-level rise.

The aggregated metrics of coastal impact analysis, as illustrated in this chapter, are helping to understand existing and future patterns of coastal vulnerability. Tools such as DIVA need to be explored in more detail and developed further, together with careful validation and verification. This should include more analysis of vulnerability across scale to develop more integrated perspectives from the individual to the global scale. A gap lies in regional analysis (e.g. Europe): tools at these sub-global scales could more explicitly link to coastal management issues, which is difficult with the current resolution of DIVA. More analysis of indirect effects would be useful, especially indirect economic effects, and wider integration with global change analysis would establish a broader context for adaptation assessment.

In conclusion, broad-scale coastal assessment offers great promise in helping to understanding and managing coastal vulnerability through the 21st century.

Acknowledgements

The authors would like to acknowledge all their colleagues on both the DINAS-COAST and Atlantis Projects, without which this paper would not be possible.

References

Arnell, N., Livermore, M. J. L., Kovats, S., Levy, P. E., Nicholls, R., Parry, M. L., & Gaffin, S. R. (2004). Climate and socio-economic scenarios for global-scale climate change impact assessments: Characterising the SRES storylines. *Global Environmental Change, 14*, 3–20.

Arrow, K. J., Cline, W. R., Maeler, K.-G., Munasinghe, M., Squitieri, R., & Stiglitz, J. E. (1996). Intertemporal equity, discounting, and economic efficiency. In: J. P. Bruce, H. Lee, & E. F. Haites (Eds), *Climate change 1995: Economic and social dimensions – contribution of working group III to the second assessment report of the intergovernmental panel on climate change*. Cambridge: Cambridge University Press.

Barnett, J., & Adger, W. N. (2003). Climate dangers and atoll countries. *Climatic Change, 61*, 321–337.

Bijlsma, L., Ehler, C. N., Klein, R. J. T., Kulshrestha, S. M., McLean, R. F., Mimura, N., Nicholls, R., Nurse, L. A., Perez Nieto, H., Stakhiv, E. Z., Turner, R. K., & Warrick, R. A. (1996). Coastal zones and small islands. In: R. T. Watson, M. C. Zinyowera, & R. H. Moss (Eds), *Impacts, adaptations and mitigation of climate change: Scientific technical analyses*. Cambridge: Cambridge University Press.

Bosello, F., Lazzarin, R., & Tol, R. S. J. (2004). *Economy-wide estimates of the implications of climate change: Sea level rise*. Research Unit Sustainability and Global Change FNU-38, Hamburg: Centre for Marine and Climate Research, Hamburg University.

Church, J. A., Gregory, J. M., Huybrechts, P., Kuhn, M., Lambeck, K., Nhuan, M. T., Qin, D., & Woodworth, P. L. (2001). Changes in sea level. In: J. T. Houghton, Y. Ding, D. J. Griggs, M. Noguer, P. J. van der Linden, & D. Xiaosu (Eds.) *Climate change 2001. The scientific basis*. Cambridge: Cambridge University Press.

Crossland, C. J., Kremer, H. H., Lindeboom, H. J., Marshall Crossland, J. I., & Le Tissier, M. D. A. (Eds). (2005). Coastal fluxes in the anthropocene. In: *The land–ocean interactions in the coastal zone project of the international geosphere–biosphere programme series: Global change – the IGBP series*. Berlin: Springer.

Darwin, R. F., & Tol, R. S. J. (2001). Estimates of the economic effects of sea level rise. *Environmental and Resource Economics, 19*(2), 113–129.

Dilley, M., Chen, R. S., Deichmann, U., Lerner-Lam, A. L., & Arnold, M. (2005). *Natural disaster hotspots: A global risk analysis*. Disaster Risk Management Series No. 5. Washington, DC: The World Bank.

Evans, E. P., Ashley, R. M., Hall, J., Penning-Rowsell, E., Saul, A., Sayers, P., Thorne, C. & Watkinson, A. (2004a). *Scientific summary: Volume I – future risks and their drivers*. London: Foresight; Future flooding, Office of Science and Technology.

Evans, E. P., Ashley, R. M., Hall, J., Penning-Rowsell, E., Sayers, P., Thorne, C., & Watkinson, W. (2004b). *Scientific summary: Volume II – Managing future risks*. London: Foresight; Future Flooding, Office of Science and Technology.

Gibbons, S. J. A., & Nicholls, R. J. (2006). Island abandonment and sea-level rise: An historical analog from the Chesapeake Bay, USA. *Global Environmental Change, 16*(1), 40–47.

Hoozemans, F. M. J., Marchand, M., & Pennekamp, H. A. (1993). *A global vulnerability analysis: Vulnerability assessment for population, coastal wetlands and rice production on a global scale* (2nd ed.). Delft Hydraulics.

IPCC CZMS. (1992). *Global climate change and the rising challenge of the sea. Report of the coastal zone management subgroup intergovernmental panel on climate change response strategies working roup*. Intergovernmental Panel on Climate Change, United States Natural, Oceanic and Atmospheric Administration, United States Environmental Protection Agency.

King, C., & Adeel, Z. (2002). Strategies for sustainable coastal management in Asia and the Pacific – perspectives from a regional initiative. *Global Environmental Change, 12*, 139–142.

Li, C. X., Fan, D. D., Deng, B., & Korotaev, V. (2004). The coasts of China and issues of sea level rise. *Journal of Coastal Research, 43*(special issue), 36–49.

Lowe, J. A., Gregory, J. M., Ridley, J., Huybrechts, P., Nicholls, R. J., & Collins, M. (2006). The role of sea-level rise and the Greenland Ice Sheet in dangerous climate change: Implications for the stabilisation of climate. In: H. J. Schellnhuber, W. Cramer, N. Nakicenovic, T. M. L. Wigley, & G. Yohe (Eds), *Avoiding dangerous climate change*. Cambridge: Cambridge University Press.

McFadden, L., Nicholls, R. J., Vafeidis, A., & Tol, R. S. J. (in press). A methodology for modelling coastal space for global assessment. *Journal of Coastal Research*.

McLean, R. F., Tsyban, A., Burkett, V., Codignotto, J. O., Forbes, D. L., Mimura, N., Beamish, R. J., Ittekkot, V. (2001). Coastal zones and marine ecosystems In: J. J. McCarthy, O. Canziani, N. A. Leary, D. J. Dokken, & K. S. White (Eds.) *Climate change 2001: Impacts, adaptation and vulnerability. Contribution of working group II to the third assessment report of the intergovernmental panel on climate change (IPCC)*. Cambridge: Cambridge University Press.

Nakicenovic, N., & Swart, R. (2000). *Special report on emissions scenarios*. A special report of working group III of the intergovernmental panel on climate change. Cambridge: Cambridge University Press.

Nicholls, R. J. (2002). Analysis of global impacts of sea-level rise: A case study of flooding. *Physics and Chemistry of the Earth, 27*, 32–42, 1455–1466.

Nicholls, R. J. (2004). Coastal flooding and wetland loss in the 21st century: Changes under the SRES climate and socio-economic scenarios. *Global Environmental Change, 14*, 69–86.

Nicholls, R. J., Hoozemans, M. J., & Marchand, M. (1999). Increasing flood risk and wetland losses due to global sea-level rise: Regional and global analyses. *Global Environmental Change, 9*, S69–S87.

Nicholls, R. J., & Klein, R. J. T. (2005). Climate change and coastal management on Europe's coast. In: J. E. Vermaat, L. Ledoux, K. Turner, & W. Salomons (Eds), *Managing European coasts: Past, present and future*. Environmental Science Monograph Series. Berlin: Springer.

Nicholls, R. J., & Lowe, J. A. (2004). Benefits of mitigation of climate change for coastal areas. *Global Environmental Change – Human and Policy Dimensions 14*(3), 229–244.

Nicholls, R. J., & Lowe, J. A. (2006). Climate stabilisation and impacts of sea-level rise. In: H. J. Schellnhuber, W. Cramer, N. Nakicenovic, T. Wigley, & G. Yohe (Eds), *Avoiding dangerous climate change*. Cambridge: Cambridge University Press.

Nicholls, R. J., & Tol, R. S. J. (2006). Impacts and responses to sea-level rise: A global analysis of the SRES scenarios over the twenty-first century. *Philosophical Transactions of the Royal Society A: Mathematical Physical and Engineering Sciences, 364*(1841), 1073–1095.

Nicholls, R. J., Tol, R. S. J., & Hall, J. W. (in press). Assessing impacts and responses to global-mean sea-level rise. In: M. Schleisinger (Ed.), *Human-Induced Climate Change: An Interdisciplinary Assessment*, Cambridge: Cambridge University Press, accepted.

Nicholls, R. J., Tol, R. S. J., & Vafeidis, A. (2005). *Global estimates of the impact of a collapse of the west Antarctic ice sheet: An application of FUND*. Research unit sustainability and global change FNU-78. Hamburg: Hamburg University and Centre for Marine and Atmospheric Science. www.uni-hamburg.de/Wiss/FB/15/Sustainability/Working_Papers.htm (last accessed 6th March 2006).

Nordhaus, W. (in press). Geography and macroeconomics: New data and new findings. *Proceedings of the National Academy of Sciences of the United States of America*.

Nurse, L., Sem, G., Hay, J. E., Suarez, A. G., Wong, P. P., Briguglio, L., & Ragoonaden, S. (2001). Small island states. In: J. J. McCarthy, O. F. Canziani, N. A. Leary, D. J. Dokken, & K. S. White (Eds), *Climate change 2001: Impacts, adaptation and vulnerability*. Cambridge: Cambridge University Press.

Oppenheimer, M., & Alley, R. B. (2004). The west Antarctic ice sheet and long term climate policy. *Climatic Change, 64*, 1–10.

Schröter, D., Polsky, D., & Patt, A. G. (2005). Assessing vulnerabilities to the effects of global change: An eight step approach. *Mitigation and Adaptation Strategies for Global Change. 10*(4), 573–595.

Small, C., & Nicholls, R. J. (2003). A global analysis of human settlement in coastal zones. *Journal of Coastal Research, 19*(3), 584–599.

Smit, B., Pilifosova, O., Burtin, I., Challenger, B., Huq, S., Klein, R. J. T., & Yohe, G. (2001). Adaptation to climate change in the context of sustainable development and equity. In: J. J. McCarthy, O. F. Canziani, N. A. Leary, D. J. Dokken, & K. S. White, (Eds), *Climate change 2001: Impacts, adaptation and vulnerability*. Cambridge: Cambridge University Press.

Tol, R. S. J. (2004). *The double trade-off between adaptation and mitigation for sea level rise: An application of FUND*. Research Unit Sustainability and Global Change FNU-48, Hamburg: Hamburg University and Centre for Marine and Atmospheric Science. http://www.uni-hamburg.de/Wiss/FB/15/Sustainability/Working_Papers.htm.

Tol, R. S. J., Bohn, M. T., Downing, T. E., Guillerminet, M.-L., Hizsnyik, E., Kasperson, R. E., Lonsdale, K., Mays, C., Nicholls, R. J., Olsthoorn, A. A., Pfeifle, G., Poumadere, M., Toth, F. L., Vafeidis, A. T., van der Werff, P. E., & Yetkiner, I. H. (2006). Adaptation to five metres of sea level rise, *Journal of Risk Analysis, 9*, 467–482.

Townend, I., & Burgess, K. (2004). Methodology for assessing the impact of climate change upon coastal defence structures. *Proceedings of 29th International Conference on Coastal Engineering* (pp. 3953–3966).

Turner, R. K., Subak, S., & Adger, N. W. (1996). Pressures, trends, and impacts in coastal zones: Interactions between socio-economic and natural systems. *Environmental Management, 20*(2), 159–173.

Vafeidis, A. T., Nicholls, R. J., Mcfadden, L., Hinkel, J., & Grashoff, P. S. (2004). Developing a global database for coastal vulnerability analysis: Design issues and challenges. In: M. O. Altan (Ed.), *The international archives of photogrammetry, remote sensing and spatial information sciences* (Vol. 35(B)). Istanbul: International Society for Photogrammetry and Remote Sensing.

Vaughan, D. G., & Spouge, J. (2002). Risk estimation of the collapse of the west Antarctic ice sheet. *Climatic Change, 52*(1), 65–91.

Yohe, G. W., & Tol, R. S. J. (2002). Indicators for social and economic coping capacity – moving towards a working definition of adaptive capacity. *Global Environmental Change – Human and Policy Dimensions, 12*(1), 25–40.

Chapter 14

Vulnerability and Beyond

Loraine McFadden, Edmund Penning-Rowsell and Robert Nicholls

Introduction

This volume has drawn together contributions from a range of disciplines to explore the challenges and opportunities for managing coastal systems, based on the analysis of the vulnerability of the coast. But we are all too aware that 'vulnerability' is a contested territory, and many different definitions often cloud issues and can prevent a clear view as to management needs and possibilities. Because of these differences we have sought to elevate the debate to a higher plane of generality by exploring, in the main, just three concepts, which can be expressed in a simple relation

Vulnerability = (**Impacts**) minus (**effects of Adaptation**)

Of course, this is an oversimplification, but it is one that we have found to be useful when moving across discipline boundaries, which is essential in vulnerability analysis, as argued by Hinkel and Klein (Chapter 5) and Schröter, Polsky, and Patt (2005) among others.

What this simple equation seeks to suggest, first, is that vulnerability of coastal systems — both human and environmental — is primarily about the propensity of those systems to be impacted by a range of drivers with adverse effects. Prime examples would be episodic events like tsunamis (Handmer, Chapter 8) or long term impacts of sea-level rise on those coastal ecosystems, which are generally adapted to present sea levels (Woodroffe, Chapter 4).

But countering impact drivers, second, is the process of adaptation. Natural systems adapt to forcing factors in many ways, in order to restore or develop some state of equilibrium. This is a process that one can see as negative feedback at work, either rapidly following some disturbing event (such as a storm) or over many years or centuries in a process of gradual adjustment and change (Woodroffe, Chapter 4). Taking the coastal ecosystems example, they can adapt by vertical accretion and horizontal migration, and sea-level rise does not then automatically translate into the adverse impacts (as net land losses), and in special cases could even translate into gains.

We hope that the equation has general application. Human coastal systems are also impacted by coastal forces and they also have an adaptive capacity, through investment in coastal defences, changes in living and livelihood patterns or locations, or in many more informal ways. Again, adaptation may be rapid, following an event that triggers human action, or can occur more slowly over time. The latter can be a gradual response to human motivations and preferences for a different relationship with coastal areas, the resources that they contain and the hazards that they can bring.

Our aspiration is for a proactive approach to vulnerability reduction. In the context of disastrous events such as the Boxing Day 2004 tsunami in the Indian Ocean and Hurricane Katrina in the southern USA there is a need for clearer thinking about how to reduce the vulnerability of our coastal areas and the populations that live there. This must be based on a clear analysis of the issues at stake, including a better understanding of the impacts that these events bring and the adaptive capacity of the coastal systems that are thereby threatened.

In this chapter, we seek a synthesis of these ideas with particular focus on key themes identified in Chapter 1 of this volume: impacts, adaptation and differences in both system behaviour and approaches for managing coastal environments and their human communities.

This synthesis inevitably raises the question of the purpose of coastal areas, how they function, and the dynamics of the balance between potential impacts and the effects of adaptation. These are difficult issues that can be answered in different ways, as this volume shows, and general management lessons do not easily emerge. However, building on the diversity of approaches that are represented here, we suggest a series of broad-scale challenges for vulnerability reduction in coastal environments and communities and this leads to some suggestions both as to lessons to learn and for future research.

A Synthesis

A range of views emerges from this volume on understanding and reducing vulnerability in coastal zone management: there are both commonalities and differences in scientific approaches to managing coastal environments.

Coastal managers generally agree that we need to take a broad and integrated view of coastal systems, so as to capture all their complexity. This is counter to most coastal science, which focuses on understanding individual system components, such as the morphology of breaker bars, the effectiveness of breakwaters or peoples' attitudes to beaches. While such understanding is vital, without integration to the system view we can miss the bigger picture that is relevant to understanding coastal vulnerability. In this way McFadden (Chapter 2) emphasises that vulnerability analysis needs a full systems perspective, with the vulnerability concept as the fulcrum around which that analysis hinges. The functional view is also generally supported by Woodroffe (Chapter 4), stressing also the importance of understanding the rate and direction of coastal changes that underpin or threaten human occupance.

But there are clear differences in basic concepts and in meanings. Hinkel and Klein (Chapter 5) advocate a formalised framework for vulnerability analysis within which coastal scientists should work, matched by a common system of communication. In contrast, Green and Penning-Rowsell (Chapter 3) see this as inherently impossible: the language of vulnerability is a function of the contested nature of visions for the coast, which emanate from stakeholders' differing 'worldviews'.

So while we seek mutual understanding in this volume, we do not seek complete convergence: different scientists see vulnerability differently, and those differences are the basis of learning from each other and moving forward.

Impacts on Vulnerable Coastal Systems

Our previous chapters have looked at a wide range of vulnerable coastal settings at a range of geographical scales. But for all, there is common recognition of the adverse impacts from hazardous events and circumstances.

Some, but not all, impacts result from climate change creating the actuality or the potential for sea-level rise, among other effects (Nicholls et al., Chapter 13). Many result from the impacts on the coast of its human occupance (Evans et al., 2004b; Tunstall & Tapsell, Chapter 7; Winchester et al., Chapter 10), especially when this occupance is intense (Jacob, Chapter 9) or intensifying (Nunn & Mimura, Chapter 12).

Most of the impacts on humans tend to be quantifiable in economic (monetary) terms. This has its expression in damage to the assets or the functions of the coastal zone, perhaps repeatedly through storm and flood damage (as with Hurricane Katrina) but also as one-off loss of the coastal area to the sea through erosion (Tunstall & Tapsell, Chapter 7).

But other human impact is more indirect. Future flooding in New York would disrupt its communication systems — as it has in the past — and thereby have an indirect impact on trade and commerce at a level that perhaps far outweighs any direct physical harm. In the same way Handmer et al. (Chapter 8) show that the informal economy of coastal Thailand was severely impacted by the Indian Ocean tsunami, and recovered slowly owing to government adaptation priorities that favoured the formal sector.

The nature and extent of impacts of forcing factors on coastal ecosystems appear quite different but in fact they are actually rather similar in that some forces are stronger and more persistent than others, have different patterns, and some ecosystems are more susceptible to harm. Complex and delicate geomorphic systems such as deltas (Sanchez-Arcilla et al., Chapter 6) that have evolved over many centuries, can be severely damaged or even lost by declining sediment fluxes: effects that are compounded by sea-level rise. The functioning of ecosystems can be disturbed by a wide range of impacts, losing productivity and ecological diversity, and the impacts can be direct or can operate indirectly by the disturbance of ecosystems dependent in some way on those suffering the direct impacts.

Physical and Human-Induced Adaptation

In discussing adaptation within the physical environment, Orford et al. (Chapter 11) highlight the finding that physical systems respond to forcing factors in ways that can — and often do — lead to a stable situation, or at least a dynamic equilibrium, that diminishes vulnerability, but that in this process sediment supply plays a major role. With sediment deficiency a coastline will react by developing new forms so as to respond to the energy of the sea. This they describe as the 'self-organising capability' of the coast, and where there is sediment 'oversupply' this process can create a myriad of new forms that again buffer the coast from the forcing factors from the sea. The 'organisation' process is one of the reaction to forces and an accommodation to those forces, over many hundreds or thousand of years. Whether a stable state is achieved will depend on whether the forcing factors themselves remain stable; if they change, then the coast is in a perpetual process of restless 'self-organisation' to seek a balance between form and process.

Human adaptation is not so very different. Impacts at the coast threaten or destroy its productive use, and hence its ability to support human populations. So these humans attempt to prevent or lessen that potentially destructive process, through a range of adaptive actions. The adaptive capacity of the coastal population will therefore determine the extent to which coastal use can remain productive, and that adaptive capacity is related to the population's ability to learn (and therefore to predict that impacts felt in the past can recur in the future) and to invest.

That investment can be in creating coast defences to prevent the sea from destroying the productive assets and uses at the coast, by enhancing the physical resistance or resilience of the coast. Or it can be in making the human use of the coast itself more resilient to attack: growing flood-resistant rice crops (Winchester et al., Chapter 10) or engineering works to protect the New York subway from floods (Jacob et al., Chapter 9). Investment means forgoing some advantage now in favour of gain in the future; this requires human judgement, and forward planning, which are distinctly human rather than physical characteristics. Proactive measures to reduce vulnerability before it is evidenced by a damaging event is perfectly possible with the modern scientific techniques at our disposal, and a type of the vulnerability analysis advocated in this volume, but it is rarely done in practise — hazardous events seem necessary to trigger policy response and action.

But human adaptation can be unplanned, just as people may rush from the threat of a tsunami (Handmer et al., Chapter 8). Communities and individuals also may make a number of small, unplanned incremental steps towards a different state, nudged along by forcing factors over which they perceive they have no means of control. Usually this will mean that the coastal zone is used in a different way, which is more resilient or upon which the people concerned have less dependence. The Islanders of the Pacific Ocean continually adjust their livelihood patterns in reaction to the changing state of the land/sea balance of forces (Nunn & Mimura, Chapter 12) and the vicissitudes of the markets for the products of their efforts.

In many developed societies the process of adaptation to coastal forces is the role of the state, not of individuals. For individuals and communities then the search for a safer and less vulnerable coast will be an indirect and political process of lobbying governments for their support to build protective measures (Tunstall & Tapsell, Chapter 7), or promote recovery from a damaging event (Handmer, Chapter 8), or for insurance arrangements or a change in land use.

Differences

The earlier chapters in this volume catalogue a range of different impacts at the coast and a wide variety of human and physical adaptations to the forces and threats that are inherent there. These 'differences' have many dimensions.

There are differences in spatial scale. The threat of coastal erosion and flooding at Corton are a pinprick on the soft east coast of England (Tunstall & Tapsell, Chapter 7), affecting a small community that is not adjusted to living on an eroding coast and which has assets that are at risk and livelihoods that are threatened. The coastal communities of

Andhra Pradesh, India (Winchester et al., Chapter 10) are much larger, and threatened with the huge forces of tropical cyclones which have caused thousands of deaths and extensive destitution in the past. The impacts of coastal hazards in the small Pacific Islands may be more limited than the previous example, but the size of these islands means that this effect can cover the whole of each island, and the whole of the population that live there.

In Andhra Pradesh, engineering solutions along thousand of miles of coast are unaffordable in impoverished India, even if they were to be effective. The process of human adaptation is therefore to build cyclone shelters and a warning system in the short term and in the longer term to build the economy of the area so that the population could then afford to protect itself as best it can in the local circumstances in which it finds itself. Scale therefore affects the economic viability of adaptation responses, and the complexity of governance arrangements.

A broad-scale perspective of the coast reveals the dominance of different coastal processes with varying spatial scales: processes at the scale of the coastal cell or the inland sea that may not be apparent at the scale of the single coastal site. Geographical scale and timescale are also interrelated (Orford et al., Chapter 11) in that slow acting geomorphological processes over geological time create new coastal forms over wide areas in response to forces that cannot easily be measured or appreciated when viewed against the small time slice that is human existence or the lifetimes of individuals. The 2004 Indian Ocean tsunami (Handmer et al., Chapter 8) affected a very large area, but the probability of its recurrence — its timeframe — is very small. The geographical extent of its impacts makes adaptation difficult for individual nations, with their limited resources, and means that the likely timescale for recovery is also large.

This leads to a further perspective on the relations between timescale and the process of adaptation, and here there are again differences across our case studies. Investment in new infrastructure in New York to protect the city from sea-level rise induced flooding has a decadal time pattern (Jacob et al., Chapter 9), although the development of a water taxi service due to a flooded rail tunnel under the Hudson River occurred almost overnight. The same decadal timeframe applies to the historical adaptation of the South Sea islanders to sea-level rise (Nunn & Mimura, Chapter 12), but the geomorphological adaptation of the Ebro delta to diminished sediment supply is barely noticeable within this timescale, but would show more dramatically over a century or more, especially with accelerated sea-level rise. Human adaptation can be much more rapid — but is not necessarily so — as coastal storms and their erosive power heightens perceptions of danger and the need for action. But there is also scale interaction — slow erosion due to sea-level rise makes future generations more vulnerable to sudden storm erosion.

Finally, there are differences in power. New York has the economic power to counter major threats from the sea (Jacob et al., Chapter 9) whereas the poor of Thailand (Handmer et al., Chapter 8) lack the capacity to help themselves in even the most minor of ways: the most vulnerable in this world are always the poor and the ill-educated. In the physical world, cyclones in the tropics have greater power than mid-latitude tidal surges such as in the North Sea, and the 'power' of a changing climate is akin to a supertanker for which a change in direction takes time and energy. The forcing factors that emanate from this changing climate are virtually unstoppable for the next half century or so whatever our future greenhouse gas emissions. Human response through greenhouse gas mitigation and adaptation has therefore to be planned over a long timescale measured in many decades, or more (Nicholls & Lowe, 2006).

Complexities and Clarity in Modelling Vulnerability through Impact and Adaptation

The above discussion highlights our conclusion that the nature of impacts within vulnerable environments and communities is relatively clear, as is the process of environmental and human adaptation. However, we can see that coastal vulnerability has many dimensions, and a range of other contributing variables needs to be considered.

The impacts that generate vulnerability affect different groups differentially. Populations or physical coastal features are more vulnerable to change if the forcing factors that lead to impacts act on systems and individuals who are inherently more susceptible to harm than others. This could be a delta compared with a rocky coast. Impacted by the same force, the former is more susceptible to change. The vulnerability of a system is therefore dependent on processes that may increase the susceptibility or resilience of the coast.

For example, deltaic subsidence, either naturally occurring or perhaps human-induced due to factors such as groundwater withdrawal, critically increases the potential for extensive impacts on the relevant coastal system. The long-term process of accelerated sediment starvation due to river regulation and damming in deltas such as the Ebro (Sanchez-Arcilla et al., Chapter 6) may decrease the overall resilience of the system. Without the sediment material to rebuild, recovery following damage now is more uncertain, if indeed it is possible at all.

In a similar fashion, when subjected to an equivalent force, the poor of Thailand and the rich of New York do not suffer the same effect, even though the absolute level of damage might be far greater in New York because the assets at risk are far more substantial. As a further example, adaptation within traditional South Sea Island communities (Nunn & Mimura, Chapter 12) may have been enhanced by living in greater 'harmony' with nature — this made them more resilient than later comers. Those living by the informal economy of Thailand (Handmer et al., Chapter 8) were less resilient than others in that their recovery process was slow and difficult.

Another important point is that adaptation is not always beneficial, and sometimes it can enhance impacts, although defining such maladaptation can be problematic as it is space and time scale dependent, and it is also wrapped up with human values and attitudes to the future (Burton, 1996). Hard defences against sea-level rise can probably cope with coastal changes in the 21st century, but will the security that they provide promote more hurricane disasters, such as Katrina, as more and more of the coastal population comes to depend on such defences. Retreat from the coast will minimise Katrina-type events, but at the expense of abandoning huge investments, forgoing the use of valuable land, and the likelihood of generating some political conflict among the evacuated populations. A middle way between these two extremes would seem prudent, but we are only beginning to understand the important choices that we face in the coastal zone.

But what is clear, even through the fog of complexity that bedevils this field, is that an assessment of vulnerability remains central to planning the use and development of coastal zones. If vulnerability is low, then there is likely to be physical stability and a good base for human occupance. High vulnerability brings the risk of damage and the risk of the need for heavy investment or abandoning the area at some future date, threatening existing assets and opportunities. This vulnerability covers multiple scales from the episodic event (the Indian Ocean tsunami; Handmer et al., Chapter 8)) to long-term pervasive change over decades and centuries (Sanchez-Arcilla et al., Chapter 6; Jacob et al., Chapter 9; Nicholls et al., Chapter 13).

Coastal areas have always had different and dynamic vulnerabilities in the past — as well as different and dynamic opportunities — but with the recognition of sea-level rise and climate change we see another force that creates the kind of long-term potential vulnerability that will change the nature of the land/sea interface, bringing mainly threats and very few beneficial opportunities. This context of increasing risks and vulnerability, coupled with the pattern of worldwide migration to the coast (Agardy et al., 2005), must serve as the background for most coastal zone management in the future.

Lessons Learnt and Challenges for Managing Coastal Vulnerability

The previous section not only highlights common themes concerning impacts and adaptation, but also the complexity of their interrelations that ultimately define the vulnerability of human and natural systems at the coast. However, while each chapter is different, cross-cutting lessons come from the convergence of ideas between chapters. Challenges also emerge from the different approaches within the volume in understanding the vulnerability of the coast.

Lessons and Challenges for Coastal Communities

Local communities are often at the 'sharp end' of vulnerable coastal environments. It is at the local level that the realities of the impact of hazardous events — and the demands and frustrations of the recovery process — are most acutely felt. Perhaps as a result, local and regional management groups and partnerships frequently provide the momentum towards real progress in coastal management (European Commission, 2001).

A simple but challenging lesson emerges from this volume, relating coastal vulnerability to community perceptions of coastal issues. This is that many coastal communities, and often those in developed countries, tend to have a somewhat idealised view of the coast, largely associated with historical and personal experiences and familiarities. This view is reflected in a desire to preserve the coast as it has been in the past (Tunstall & Penning-Rowsell, 1998). It conceptualises the coast as a benign environment (Tunstall & Tapsell, Chapter 7), or at least one that can be (and should be) controlled, contained and maintained, to ensure physical as well as socio-economic stability.

In turn this perception governs local intervention and lobbying that strongly advocates 'holding the line', which often in fact can have longer-term destabilising effects on the physical coastal system rather than the reverse. Physical systems need space to adjust to the forces that mould them, and if coastal management can promote coastal buffers to provide this space, rather than squeezing the shoreline into an outdated form, this could have important benefits in reducing coastal vulnerability (see also Rochelle-Newall et al., 2005).

Tackling such discontinuity between the relatively static human constructs about the coast and the dynamics of physical coastal environments has characterised coastal management since its earliest days. The 'conservative' nature of many coastal communities, for example, means that widening the scope of — or enhancing — the adaptive capacity of coastal systems is often frustrated. Deep-rooted and slowly changing societal constructions of 'the coast' therefore pose a particular challenge to coastal managers. Progress towards further changing such ideas with modern integrated coastal zone management (ICZM) concepts could have significant benefits in enhancing the capacity for sensible adaptation.

The applicability of such modernised 'social reconstruction' of the coast can be seen in relation to the New Orleans coastal community and other such regions which suffer from particularly damaging but not unexpected events (Jonkman, Stive, & Vrijing, 2005; Pilkey & Young, 2005). A careful deconstruction of New Orleans (assessing what can or needs to be preserved and retreating or replanning the city where possible) will have a greater impact on the vulnerability of the region than simply reconstructing the city as it was before (Jacob, 2005). Such an approach would provide a more extensive and long-term improvement in the ability of the region to withstand the increasing external forcing that the next century of sea-level rise is likely to bring. However, it requires communities and community leaders to cultivate a series of different perceptions and expectations of coastal environments.

This highlights another simple fact. Hard decisions need to be made in both short- and long-term management of coastal zones despite the complex interactions between various subsystems, which make such decisions particularly difficult. Given this complexity, stakeholders (including and especially local communities and coastal managers) need to foster a long-term vision of the coast against which to plan both the immediate and the future development of the system.

This may result in having to make difficult but fundamental decisions in order to prescribe the long-term physical, social and economic goals for the region. Agreeing with such goals would be a demanding and perhaps sometimes impossible process, given the inherent conflicts at the coast (Green & Penning-Rowsell, Chapter 3). However, by linking with the scientific community, stakeholders can move towards developing a longer-term perspective of the functioning system, and work towards recognising and achieving desirable and 'stable' future states and 'goals' for the coast. Vulnerability analysis certainly contributes in this way to understanding current and potential future impacts and responses of the coast to external perturbation. As suggested by McFadden (Chapter 2), we can build on this understanding to create environments that sustain or enhance the physical and socio-economic functions that coastal zones provide.

Lessons and Challenges for the Process of Managing the Coast

Like the coast itself, coastal vulnerability can be considered a dynamic phenomenon. Coastal processes — physical and human — change through space and with time, and these changes define the differential adaptive responses of coastal society or the physical environment that are needed to match external forcing. A basic lesson for ICZM from this volume is that management decisions should be based on a greater understanding of the coast in a dynamic and total-system framework.

But this total-system approach is likely to remain a challenge, particularly if the responsibility for implementing ICZM is focused at regional and local levels, such as in the European institutional context. It is important to ensure that scientific approaches and models, contributing to our understanding of the integrated nature of coastal behaviour at multiple scales, are available to support all levels of decision making for integrated management. While both physical and social scientists have developed a fundamental understanding of the behaviour of many components of the coastal system, integrating this knowledge to understand the total behaviour of the coast and applying this knowledge to achieve tangible outcomes for coastal management are still major challenges.

The fact that there are wide ranges of spatial and temporal scales in both impact and adaptation within coastal systems also suggest an important implication for management. By intervening within the coast at any point in space or in time, our actions can potentially have wide-ranging effects. Activities within coastal systems now can affect future impacts and adaptive or coping capacities; this highlights the centrality of developing the scientific knowledge basis of ICZM programmes and strategies. There are many unknowns surrounding the relationships between the physical and human environment and feedback mechanisms and scaling issues (behavioural links across space and in time). This means at present it remains difficult to anticipate, and therefore manage, the directions and magnitudes of change in those thresholds, which define the actual vulnerabilities of the coast.

In fact, the stark reality may be that due to past misguided intervention within coastal systems, vulnerability thresholds may already have been crossed which limit the capacity to achieve integrated management of the coast. An interesting example is given by Orford and his colleagues in highlighting the impact of the accumulation of centuries of forcing disequilibrium in the geomorphic environment by coastal engineering and reclamation on the UK coastline (Orford et al., Chapter 11). Sediment pathways have broken down or are in deficit mode and, given this, there is a loss of capacity of an important component of the coastal system.

A general lesson may be drawn from this inheritance of coastal management: unless we invigorate the dynamics and so the health of the coastal system, we may seriously limit our capacity to take a strategic approach to coping with pressures on the coastal zone. This also relates to human dimensions of impact and recovery from external forcing events: when elements of society are in a depleted condition (e.g., in terms of human health, wealth and access to resources), the capacity for a comprehensive and strategic adaptive response may be greatly reduced. The recovery process associated with the Indian Ocean tsunami (Handmer et al., Chapter 8) and also that of Hurricane Katrina are examples of this problem, as are the difficulties of coastal management in impoverished India (Winchester et al., Chapter 10).

An important lesson also concerns the importance of managing the land and water interface together. Furthering our understanding of the interactions between the hinterland and coast, and examining the role of boundary effects in determining change in the coastal system, may provide important insights for reducing impacts and enhancing the adaptive response to system perturbations.

However, it is important to note that at this interface, vulnerability is also characterised by a complexity of institutional frameworks. It is very difficult to coordinate management across the wide range of land-water divides. An important lesson emerging from this discussion is that while integration is central to effective management, we may not (and often cannot) have the 'luxury' of a single ICZM body to guide decision making and the management process. While there are countries in which the coast is a national priority (e.g., the Dutch perspective on their coastal environment), this is atypical and as such it is naïve to believe that such national emphasis can be achieved in all coastal regions. Given this reality, effort must be placed on facilitating broad-based decision making, including enabling the appropriate governance and science-based mechanisms to work together to allow a more comprehensive approach to managing vulnerable coastal systems than would otherwise be the case.

Lessons and Challenges for Policy Makers

We have discussed a wide range of options for reducing vulnerability through coastal management. However the material within this volume also draws attention to a number of limitations to the use of vulnerability analysis. These limitations raise particular challenges and lessons for decision makers in coastal management.

This volume suggests that a critical concept in managing the coast is that of 'lag time' in system behaviour. The lag in a system is the time interval between an external or internal forcing event, and the significant impact of that event on the characteristics and behaviour of the coast. This is especially important when different responses are nested (Woodroffe, Chapter 4; see also, Stive et al., 2002). A lagged reaction within the system may present significant challenges to the effectiveness of policy decisions in coastal management because it means that policy action and system reaction may well not coincide, which may result in uncertainty as to the most effective strategy or set of strategies to mitigate or to absorb pressures. Or the effect of a forcing factor such as climate change may not show up for decades or centuries, and may persist for many years or decades after the forcing agent has long since dissipated.

A human perspective on lag time is that related to the adaptive response of the system. This dimension reflects the delayed time between a policy being implemented and the full adaptive response in the system. Handmer and his colleagues highlight such behaviour, showing vulnerability to the impacts of the Indian Ocean tsunami being exacerbated in Phuket, Thailand, as a result of delayed government intervention (Handmer et al., Chapter 8). Coastal protection in England (Tunstall & Tapsell, Chapter 7) is another example of a lagged process of adaptive response. The East Coast of England flooding of 1953 stimulated a process of developing evidence-based standards of defence for the coastline, with attention focusing particularly on the Thames Estuary region. Further research and eventually legislation provided the basis for the construction of the Thames Barrier: however, this facility only became operational some 30 years later. It is partly down to chance that London did not experience a major flood in this 30-year window, and on several occasions water was lapping at the top of the old defences.

An important issue therefore emerges from this volume. Delays in the full implementation of adaptive responses on the coast may create windows of time throughout which environments and communities retain high levels of vulnerability, which is very difficult to counter despite moves towards adaptation. One important lesson that can be drawn from this discussion is that decision makers may not be able persistently to delay strategy development and implementation, while waiting for increasing levels of certainty concerning the nature of forcing agents on our coasts (climate change being a prime example). As decisions on management options are postponed, so the existing vulnerability of the system is also extended further into the future.

However, while it is relatively easy to discuss best approaches to adaptive management at an abstract level, a basic point is that adaptation, and indeed coastal zone management in general, can be an expensive business. The chapters in this volume that focus on coastal vulnerabilities in developing countries emphasise that the distribution of resources — the political economy — is a critical factor in determining the vulnerability of individuals and communities. In many such regions the effectiveness of vulnerability analysis as a management tool is based on the range of cheap and sustainable options that are available to policy makers to address critical issues that the analysis may identify. In combination with such appropriate policy options, individuals, households and communities themselves require a sufficient resource base upon which to respond to policies and best approaches for management.

Such requirements lead to a further challenge in terms of managing vulnerability within coastal systems. Individuals and coastal communities are vulnerable to a wide range of hazards, not just those which are coastal-specific. To be most effective in reducing vulnerability, our understanding of the range of actual and potential impacts of external forcing, and approaches to enhancing adaptive capacity, should be based within a wide context of societal, physical and economic perturbations and responses. If conditions are created that support sustainable development in socio-economic and natural systems, while providing individuals and communities with an acceptable quality of life, then the capacity of communities to adapt to a wide range of external forcing factors will be greatly enhanced. The trauma and widespread loss of functionality within communities along the Louisiana, Mississippi and Alabama coastlines, given Hurricane Katrina, may demonstrate this. Although a hurricane was the stimulus of the tragedy, the base of the disaster was 'human' occupance and behaviour. Of course, alleviating poverty will not solve all the problems of vulnerable coastal zones: the range of lessons and challenges emerging from this volume indicates otherwise. However, enhancing the functional capacity of the poorest may greatly increase the resilience of many such communities.

Research Needs: Enhancing the Evidence Base

Understanding the vulnerability of coastal systems can increase knowledge about the total behaviour of the coast, encouraging management policies to be focused around a broad range of spatial and temporal scales. It can identify dimensions of coastal environments that need particular targeting and can act as an integrating vehicle in a context where complexity and conflicts of interests are key features, both between and within physical and socio-economic environments. In such respects, vulnerability analysis can act as a useful tool for coastal management.

There are limitations in the use of vulnerability analysis, and specific recommendations for managing coastal systems depend on quite challenging and comprehensive changes to our use and perception of coastal systems. However there is a simple message that comes from this volume: the best prescriptions for vulnerability reduction are those which are most strongly evidence based. Increasing this evidence base must therefore be a research priority.

Research Needs in the Social Science Domain

The agenda for social science research in vulnerability analysis is potentially large. However, the five key areas may be identified (Table 14.1) that would repay further detailed analysis across a spectrum of coastal scales, and not just with the traditional single case study approach.

Given that there is data on the impacts of coastal hazards — although it is patchy (Nicholls et al., Chapter 13) — one area on which to concentrate is the process of adaptation (Table 14.1). Underpinning the effectiveness of this process is the adaptive capacity of individuals, agencies and indeed nations. Adaptation is closely related to the resources, commitment and 'vision' of stakeholders at a range of scales. Understanding adaptive capacity across such scales would give a clear perspective of the most effective balance between community-driven and strategic regional approaches to managing the coast.

Table 14.1: Five key areas where coastal vulnerability research in the social sciences would produce valuable results.

Research topic area	The research questions that need attention	How might this be achieved?
1. Adaptation: the role of individuals and the state	How far can we rely on individuals and local communities to help themselves?	Longitudinal studies of communities that suffer from hazards and then adapt successfully
2. Adaptive capacity	What factors promote the enhancement of adaptive capacity in coastal communities or regional agencies?	A comparison of communities or agencies that have and have not appeared to adapt successfully to the coastal hazards that they face
3. Post-event recovery	What, in addition to direct 'outside' aid, promotes or inhibits the recovery from coastal hazards and how can insight here be used to enhance recovery processes?	In depth study of recovery timelines (e.g. Hurricane Katrina; the Indian Ocean tsunami), with systematic study of all major coastal disasters
4. Coastal resources and human attitudes	What are the fundamental motivational forces for human coastal occupance, and what affects the strength of those forces?	Substantial socials surveys in different countries and contrasting vulnerability circumstances
5. Governance issues	What is the effect of different governance structures on the efficacy of vulnerability reduction?	Institutional analysis of different vulnerability situations, probably in different countries

We know little about how this capacity might be enhanced, because we know very little about the factors that inhibit or promote this capacity across this spectrum of scales. Future research should address this and include the short- and long-term processes of recovery from hazardous coastal events, as a key adaptive behaviour.

More fundamental, and in the 'deep background' behind much vulnerability analysis, are human attitudes towards the use of coastal resources, and human aspirations for their enhancement and development. We know in the UK that humans have a reactionary streak here — seeking a continuation of past experiences and 'Victorian' associations (Tunstall & Penning-Rowsell, 1998). However, we know little of such ideas and trends in other countries or in different cultural circumstances.

Coastal human vulnerability derives from what are, on the face of it, unwise patterns of human occupancy. Our understanding is poor as to what drives these human patterns, and whether they will endure. This is more sociological than economic, and concerned with the way people see their locus in the world, and what they expect of that environment in terms of risks and rewards.

To be most rewarding this research direction needs to be matched by research into different governance arrangements. This would include assessing the needs and opportunities for implementing better coastal policies, and the effects of different decision-making processes on vulnerability reduction.

Research Needs in the Environmental/Engineering Science Domain

The environmental science and engineering components of coastal vulnerability are widely researched, but not necessarily for coastal vulnerability assessment.

The process revolution of the 1960s ended much of the typological investigation of natural phenomena and focussed on small-scale process studies. While global change issues and also technology (remote sensing, GIS, etc.) have caused a partial swing of the pendulum back towards broad-scale studies, new research vigour is required at these broad scales. As an example, the DIVA model was forced to use global coastal datasets from the 1950s (McFadden, Nicholls, Vafeidis, & Tol, in press), because nothing better has been produced in the intervening period, despite the scope for massive improvement. Table 14.2 focuses on four areas where research investment in environmental/engineering sciences would provide most benefit to understanding of coastal vulnerability. Most of the suggestions are broad-scale research reflecting the weakness identified above.

Regional-to-global coastal typologies could be greatly improved using new global datasets such as SRT digital elevation model from the Shuttle Mission with existing data sets (e.g., geology) and remote sensed data (e.g., geomorphology). This could be extended to include all coastal factors such as coastal population and hence become an integrated science activity (cf. McFadden et al., in press).

The overall response of the coast to sea-level rise, climate change and other possible drivers (e.g., declining sediment budgets) remains poorly understood. The integrated approach of Cowell et al. (2003a, 2003b) provides a good basis for understanding broad-scale sediment budgets and hence changes in shoreline position. These types of methods should be developed further and applied widely to a wide range of coastal types (see also Dickson, Walkden, Hall, Pearson, & Rees, 2005).

Table 14.2: Understanding coastal vulnerability: Four key research topics in the environmental/engineering science areas.

Research topic area	The research questions that need attention	How might this be achieved?
1. Coastal typology	How is the world's coast best classified for vulnerability analysis from a physical basis?	Use of new global datasets to improve existing typologies
2. Overall coastal change	How will the coast respond to sea-level rise and other drivers?	Build on the 'coastal tract cascade' concept
3. Ecosystem response	How will coastal ecosystems such as saltmarshes respond to sea-level rise and other drivers?	Build on existing landscape models
4. Soft engineering	What is the potential for soft engineering to counter predicted coastal recession?	Examine the likely costs and constraints over the long-term, and hence determine potential for application under different change scenarios

Broad-scale ecosystem response to climate change and other drivers requires particular attention due to the importance of these systems and their complex potential responses. Landscape models have been developed in a few regions (e.g., the Mississippi delta (Reyes et al., 2000; Martin, Reyes, Kemp, Mashriqui, & Day, 2002)). This approach needs to be applied more widely, combined with remote sensing for ground-truth.

Understanding the costs and constraints of engineering responses is a key element in the analysis of coastal vulnerability. This can be usefully separated into an assessment of soft and hard engineering within the context of integrated assessment modelling for coastal vulnerability assessment.

Research Needs in the Integrated Coastal Science Domain

The conceptual basis and implementation of integration still requires significant development. This is critical in terms of moving from an academic understanding of coastal vulnerability to operational tools that support coastal management and related areas of policy (e.g., climate mitigation policy). Table 14.3 focuses on four areas where research investment in integrated assessment would improve our understanding of coastal vulnerability and its application.

Scaling and nesting issues are fundamental to linking studies through space and time. Links could be formal such that local studies are 'nested' in a broad-scale assessment, which provides context and consistency to more local studies, and hence the overall

Table 14.3: Understanding coastal vulnerability: Four key research topics in integrated science.

Research topic area	The research questions that need attention	How might this be achieved?
1. Understanding nesting and scaling issues	How can one best integrate between adjacent scales?	Exploration of linkage between existing global to sub-national assessment tools and local scale
2. Develop integrated frameworks for coastal vulnerability	How are different disciplines best integrated in coastal vulnerability studies?	A combination of vulnerability theory and testing in application.
3. Moving from climate change to models of all drivers of change	How can we move from models of single drivers or issues to the more general models that coastal policy and management needs?	Building on the existing broad-scale models of climate change, other change factors can be added as appropriate
4. Characterising the consequences of adaptation	What are the benefits and costs of adaptation over a range of time and space scales?	Consider adaptation from a range of perspectives (economic, environmental and human) over a significant time span (50–100+years)

assessment operates at multiple scales. To date, there is limited experience of this approach for coastal vulnerability but the need for such research is becoming widely acknowledged, building on the growing recognition of the importance of understanding the multiple dimensions of coastal behaviour. Integrated frameworks for coastal vulnerability assessment are again fundamental and require ongoing development. There are emerging frameworks such as provided by Hinkel and Klein (Chapter 5) and Schröter et al. (2005). Widespread empirical testing of their effectiveness is required.

However, there seems to be no (or limited) link between this work and current progress within the coastal management community, which is largely focused on governance issues. The utility of vulnerability analysis could be improved with a clearer understanding of the challenges in linking integrated science and models of good governance (McFadden, submitted).

Many vulnerability assessments focus on single issues such as climate change, although these are often complex in themselves. For coastal management as opposed to other purposes, an assessment of all the drivers of change is required. The Foresight Future Flooding study represents a good example of a national scale study (Evans et al., 2004a, 2004b), but this needs to be extended across scales — to the local and the global.

Lastly, the consequences of adaptation are a product of the integrated coastal system, including interaction with all the physical and human drivers. It is important strategically to explore the benefits and costs of adaptation using a range of measures and a range of time and space scales up to 100+ years. This will allow the identification of possible maladaptation and hence promote debate about the diverse adaptation options available. Evans et al. (2004d) provides a good example of one approach to this goal, which could be repeated using other tools, or tools could be developed explicitly for this purpose.

Conclusions

Coastal vulnerability is an integrating concept. The diverse group of people involved in this book — engineers, geographers, geomorphologists, economists, sociologists, ecologists, geologists/geophysicists — all embraced the concept of vulnerability, its value to society, and especially as a way of making their knowledge more relevant to coastal managers and stakeholders.

However, different disciplines see vulnerability differently, reflecting the mindset and culture of each. So as to understand coastal vulnerability and hence have the capacity to manage it, all these different views need to be appreciated. This integration is happening to some degree, compared to a decade ago, but it still requires further effort to be effective.

Vulnerability has meaning at a range of time and space scales from the individual and local to the national and global. It is meaningful to explore methods for interpreting and managing vulnerability across this wide range of scales, so long as the questions that are posed are appropriate for that scale — for example, a global study can tackle global questions, but cannot be expected to resolve sub-regional issues. We take a broader view of vulnerability than some authors (e.g., Schröter et al., 2005), because we see regional or even local issues always best viewed in their wider context, and we believe that this is both theoretically meaningful and policy relevant to reducing coastal vulnerability.

Vulnerability is a product of human decisions, so coastal vulnerability can be managed by taking appropriate — and alternative — decisions. A simple model such as that underpinning this volume [**Vulnerability** = (**Impacts**) minus (**effects of Adaptation**)] is attractive as it captures the absolutely essential points. However, to operationalise the concept of coastal vulnerability for management requires more complex conceptual viewpoints as referenced, for example, by McFadden (Chapter 2), Hinkel and Klein (Chapter 5), among others.

Adaptation can profoundly modify vulnerability: this large capacity needs to be exploited positively throughout the 21st century if we are not to see a large growth in coastal vulnerability. However, not all adaptation achieves its goal of reducing vulnerability — sometimes vulnerability rises due to maladaptation. Recognising maladaptation depends on timescale — flood defences may reduce short-term vulnerability, but promote a longer-term rise in vulnerability as population and assets continue to increase behind the defences.

Does Hurricane Katrina indicate maladaptation in New Orleans, and in London will we look back on the Thames Barrier (Gilbert & Horner, 1984) and the Thames 2100 project (Lavery & Donovan, 2005) as maladaptation? To some the answer is self-evidently 'yes', but this view is by no means universal and some might think that New Orleans could be rebuilt to be better than it was before the hurricane, with the flood being a blip on a positive trajectory. Hence, the issue of maladaptation needs to be further explored within integrated analysis frameworks.

Related to maladaptation, successful adaptation in one sector (e.g., human safety) can greatly increase the vulnerability of other system elements (e.g., coastal squeeze of coastal ecosystems). Only an integrated analysis can see these tradeoffs and lead to more optimal adaptation at the coastal system level. The Regional Impact Study (RegIS) (Holman et al., 2005a, 2005b) and the Foresight Study of Flood and coastal defence exemplify this level of study (Evans et al., 2004c, 2004d).

Vulnerability can be reduced, but it can never be removed, despite the seemingly widespread human desire for a risk-free world. This point may seem self-evident, but it is worth reminding non-specialists of the limits to managing vulnerability. As hazardous areas are attractive to human populations because they contain resources that benefit those people (e.g., Burton, Kates, & White, 1993), so we must accept that tolerating some degree of vulnerability can provide profound benefits to those who accept these risks. Our goal is to make sure that the benefits outweigh the costs and to ensure that when 'things go wrong' the afflicted coastal society has the capacity to cope with the adverse situation, rather than fail to do so.

References

Agardy, T., Alder, J., Dayton, P., Curran, S., Kitchingman, A., Wilson, M., Alessandro, C., Restrepo, J., Birkeland, C., Blaber, S., Saifullah, S., Branch, G., Boersma, D., Nixon, S., Dugan, P., Davidson, N., & Vörösmarty, C. (2005). Coastal systems. In: R. Hassan, R. Scholes, & N. Ashe (Eds), *Ecosystem and human well-being: Current state and trends, Volume 1: Findings of the Condition and Trends Working Group*. USA: Island Press (Millennium Ecosystem Assessment Series).

Burton, I. (1996). The growth of adaptation capacity: Practice and policy. In: J. D. Smith, N. Bhatti, G. Menzhulin, R. Benioff, M. Budyko, R. Campos, B. Jallow, & F. Rijsberman (Eds), *Adapting to climate change*. New York: Springer.

Burton, I., Kates, R. W., & White, G. F. (1993). *The environment as hazard* (2nd ed.). New York: The Guilford Press.

Cowell, P. J., Stive, M. J. F., Niedoroda, A. W., De Vriend, H. J., Swift, D. J. P., Kaminsky, G. M., & Capobianco, M. (2003a). The coastal tract (part 1): A conceptual approach to aggregated modelling of low-order coastal change. *Journal of Coastal Research*, *19*(4), 812–827.

Cowell, P. J., Stive, M. J. F., Niedoroda, A. W., Swift, D. J. P., De Vriend, H. J., Buijsman, M. C., Nicholls, R. J., Roy, P. S., Kaminsky, G. M., Cleveringa, J., Reed, C. W., & De Boer, P. L. (2003b). The coastal tract (part 2): Applications of aggregated modelling of lower-order coastal change. *Journal of Coastal Research*, *19*(4), 828–848.

Dickson, M. E., Walkden, M. J. A., Hall, J. W., Pearson, S., & Rees, J. (2005). Numerical modeling of potential climate change impacts on rates of soft cliff recession, northeast Norfolk, UK. *Proceedings of coastal dynamics '05*, 4–8 April 2005, Barcelona, Spain, American Society of Civil Engineers.

European Commission. (2001). EU focus on coastal zones. Turning the tide for Europe's coastal zones. Directorate-General Environment, Nuclear Safety and Civil Protection.

Evans, E., Ashley, R., Hall, J., Penning-Rowsell, E., Sayers, P., Thorne, C., & Watkinson, W. (2004a). *Scientific summary: Volume II. Managing future risks. Foresight; future flooding*. London: Office of Science and Technology.

Evans, E. P., Ashley, R. M., Hall, J., Penning-Rowsell, E., Saul, A., Sayers, P., Thorne, C., & Watkinson, A. (2004b). *Scientific summary: Volume I — future risks and their drivers. Foresight; future flooding*. London: Office of Science and Technology.

Evans, E. P., Ashley, R. M., Hall, J., Penning-Rowsell, E., Saul, A., Sayers, P., Thorne, C., & Watkinson, A. (2004c). *Scientific summary: Volume I — future risks and their drivers. Foresight; future flooding*. London: Office of Science and Technology.

Evans, E. P., Ashley, R. M., Hall, J., Penning-Rowsell, E., Sayers, P., Thorne, C., & Watkinson, W. (2004d). *Scientific summary: Volume II. Managing future risks. Foresight; future flooding*. London: Office of Science and Technology.

Gilbert, S., & Horner, R. (1984). *The Thames Barrier*. London: Thomas Telford.

Holman, I. P., Nicholls, R. J., Berry, P. M., Harrison, P. A., Audsley, E., Shackley, S., & Rounsevell, M. D. A. (2005a). A regional, multi-sectoral and integrated assessment of the impacts of climate and socio-economic change in the UK: II results. *Climatic Change, 71*(1–2), 43–73.

Holman, I. P., Rounsevell, M. D. A., Shackley, S., Harrison, P. A., Nicholls, R. J., Berry, P. M., & Audsley, E. (2005b). A regional, multi-sectoral and integrated assessment of the impacts of climate and socio-economic change in the UK: I methodology. *Climatic Change, 71*(1–2), 9–41.

Jacob, K. (2005). Time for a tough question: Why rebuild? 6 September 2005. *Washington Post*. Available online- http://www.washingtonpost.com/wp-dyn/content/article/2005/09/05/AR2005090501034.html (last accessed 2nd March 206).

Jonkman, S. N., Stive, M. J. F., & Vrijing, J. K. (2005). New Orleans is a lesson to the Dutch. Editorial. *Journal of Coastal Research, 2*(6), xi–xii.

Lavery, S., & Donovan, B. (2005). Flood risk management in the Thames Estuary looking ahead 100 years. *Philosophical Transactions of the Royal Society, A363*, 1455–1474.

Martin, J. F., Reyes, E., Kemp, G. P., Mashriqui, H., & Day, J. W. (2002). Landscape modelling of the Missippi delta. *Bioscience, 54*, 357–365.

McFadden, L. (in press). Governing coastal spaces: The case of disappearing science in Integrated Coastal Zone Management. *Journal of Ocean and Coastal Management*.

McFadden, L., Nicholls, R. J., Vafeidis, A., & Tol, R. S. J. (in press). A methodology for modelling coastal space for global assessment. *Journal of Coastal Research*.

Nicholls, R. J., & Lowe, J. A. (2006). Climate stabilisation and impacts of sea-level rise. In: H. J. Schellnhuber, W. Cramer, N. Nakicenovic, T. Wigley, & G. Yohe (Eds), *Avoiding dangerous climate change*. Cambridge: Cambridge University Press.

Pilkey, O. H., & Young, R. S. (2005). Will Hurricane Katrina impact shoreline management? Here's why it should. *Journal of Coastal Research, 2*(6), iii–x.

Reyes, E., White, J. L., Martin, J. F., Kemp, G. P., Day, J. W., & Aravamuthan, W. (2000). Landscape modeling of coastal habit change in the Mississippi delta. *Ecology, 81*(8), 2331–2349.

Rochelle-Newall, E., Klein, R. J. T., Nicholls, R. J., Barrett, K., Behrendt, H., Bresser, T. H. M., Cieslak, A., de Bruin, E. F. L. M., Edwards, T., Herman, P. M. J., Lanne, R. P. W. M., Ledoux, L., Lindeboom, H., Lise, W., Moncheva, S., Moschella, P. S., Stive, M. J. F., & Vermaat, J. E. (2005). Global change and the European coast: Climate change and economic development. In: J. E. Vermatt, Bouwer, L., Turner, R. K., & Salmons, W. (Eds), *Managing European coasts: Past, present and future*. New York: Springer, (Environmental Science Monograph Series).

Schröter, D., Polsky, D., & Patt, A. G. (2005). Assessing vulnerabilities to the effects of global change: An eight step approach. *Mitigation and Adaptation Strategies for Global Change, 10*(4), 573–595.

Stive, M. J. F., Aarninkoff, S. J. C., Hamm, L., Hanson, H., Larson, M., Wijnberg, K., Nicholls, R. J., & Capobianco, M. (2002). Variability of shore and shoreline evolution. *Coastal Engineering, 47*, 211–235.

Tunstall, S., & Penning-Rowsell, E. (1998). The English Beach: Experiences and values. *The Geographical Journal, 164*(3), 319–332.

Subject Index

adaptation, 3–7, 223–229, 232–235, 237–238
adaptive capacity, 4–7, 10–11, 13, 16, 21, 42, 56, 61–62, 66, 115, 225, 244, 246, 249, 253–254
agriculture, 85, 90, 93, 162, 167, 171, 173, 201, 205, 207, 216–217

beach, 46–49, 51–55, 87–88, 90–91, 104–105, 107–115, 141, 143, 145–146, 149, 151, 153, 155, 180–182, 185, 187–190, 201–205, 209, 217

challenges, 243–244, 249–253, 257
choice, 29–34, 40
climate change, 61–64, 75, 141, 144–145, 148, 151, 155–156, 214–219
coastal change, 16–18, 21, 24, 26
coastal defences, 107–109, 179–181, 183, 187, 192, 244
coastal functions, 24, 81, 86, 87, 88, 89, 93, 178, 217, 234, 250
coastal hazard, 141–142, 150–151
coastal morphodynamics, 50, 81, 83, 93
coastal protection options, 104
coastal wetlands, 87
communication, 30, 37–38, 62, 67, 74, 152, 244
conflict, 30–31, 34, 42
coral reefs, 56, 58–59, 198, 201, 206, 216–218
cyclones, 159–165, 167, 169, 173, 214–215

delta, 52, 79–93, 160, 162, 171, 173, 190, 198, 199, 235–237, 247–248, 256
delta reshaping, 79

disaster, 2, 123, 125–126, 129–133, 135–136, 150, 152, 167, 169, 217–218

equilibrium, 46–49, 52, 55–56, 82–84
erosion, 83, 85–86, 90, 97, 102, 106, 184–187
estuary, 52–53, 102, 151, 182, 189–190, 235, 252
exposure, 6, 161, 178, 230, 232–233, 236
extreme events, 147, 160, 163–164, 211

Foresight Future Flooding and Coastal Defence Project, 19, 190, 225, 236
flooding, 97–102, 112–113, 141–156, 159–162, 170, 230–234, 236
fragility, 4, 6, 150

geomorphology, 49, 51
government, 102, 104, 111–114, 118, 125–126, 128–129, 134, 136, 150, 152, 156, 160–161, 163–165, 167–169, 174, 197, 209, 211, 213–214, 218

habitat, 23, 45, 56, 85, 87, 142, 146, 192
health, 97–98, 101, 126, 131, 164, 165
household, 35, 39–41, 99, 101, 134, 164, 169, 173, 225–226

impacts, 1–3, 5–7
informal economy, 125, 127, 134–135
infrastructure, 19, 90, 92, 97–98, 126, 129, 132, 141, 147, 150–153, 155, 165, 167, 174, 205, 211, 213, 218, 247
insurance, 102, 130, 132, 147–148, 150–151, 154, 246

integrated assessment, 4, 9, 10, 17, 22, 23, 24, 26, 67, 70, 74, 227, 256
Integrated Coastal Zone Management, 8, 26, 169, 217, 250
IPCC, 15, 18, 61, 63–64, 71–72
islands, 195, 197–198, 201, 203–209, 211–219, 235–237

knowledge integration, 62, 67, 68, 74

land use, 151, 152, 153, 169–173
language, 30, 34–35, 37–38, 41–42
lessons, 244, 249–250, 252–253
livelihoods, 121–122, 124–125, 127, 129, 132–133, 135–136
local communities, 97, 112, 116, 118, 129–130, 133–135, 197, 211, 213–215, 218
local economy, 121, 124–126, 129, 132, 136

maladaptation, 236, 248, 258–259
mangroves, 202, 206, 212, 216–217
metrics, 3, 10, 21–22, 224, 226, 229, 238
mitigation, 150–152, 156, 223–225, 230, 235, 237
modern management, 209

natural systems, 177
NGOs, 125, 128, 134, 136, 161, 168, 212

options, 30–33, 97, 104, 108–110

policy change, 102, 104, 112–113, 116
political economy, 159, 163–165, 167, 169, 173

realignment, 97, 102, 112–113, 115–118, 189–190, 237
recovery, 121–122, 124–126, 129–136
research needs, 253–256
reclamation, 182, 187, 190, 215–216

resilience, 4–7, 17, 21, 45–46, 49, 51, 55–56, 58, 121–122, 127, 134–136, 178–192
resources, 121, 127–128, 134, 136, 159, 161, 163–166, 173
risk, 38, 40, 97, 102, 104, 112, 118, 130, 141, 148, 152, 164–167 170

scale, 20–21, 50–52, 58, 224–226, 228–231, 235, 237–238
scenarios, 20, 57, 64–65, 69, 71–74, 143–145, 155, 169–171, 173, 179, 190–192, 215, 224, 227–228, 230, 232–233, 234–236, 256
sea–level rise, 62–65, 70, 72–73, 75, 92, 141, 144–145, 223–230, 232–237
sediment, 18, 23, 46, 48, 50, 52–53, 56, 79, 81–85, 87, 92, 102, 107, 177–192, 198, 203–204, 207, 209, 216–217, 245, 247–248, 251, 255
sediment husbandry, 179, 189–190
segmentation, 186–188
self–organisation, 46, 52, 56, 180, 184, 191
sensitivity, 16, 23, 46, 61, 72, 177
stakeholder, 26, 33, 152, 154
storm surge, 141, 143–145, 147, 150–152, 155, 213, 215, 217, 230
system, 22, 35, 37, 39–41
susceptibility, 4, 6, 46, 164, 178–180, 190–192, 248

thresholds, 50, 52, 58–59, 180
tourism, 85, 121–124, 126–135, 215, 216
traditional management, 208, 219
trans–disciplinary, 17–18
tsunami, 121–124, 128–133, 135–137, 213
typology, 38, 39, 228

uncertainty, 30, 32, 104, 159, 165, 167

warning(s), 112, 121, 124, 131–132, 161, 167, 217, 247